Fluorides and Dental Fluorosis

Monographs in
Oral Science

Vol. 7

Editor
Howard M. Myers, Philadelphia, Pa.

S. Karger · Basel · München · Paris · London · New York · Sydney

Fluorides and Dental Fluorosis

HOWARD M. MYERS
School of Dental Medicine, Center for Oral Health Research,
University of Pennsylvania, Philadelphia, Pa.

2 figures and 15 tables, 1978

S. Karger · Basel · München · Paris · London · New York · Sydney

Monographs in Oral Science

Vol. 3: CIMASONI, G. (Geneva): The Crevicular Fluid.
VIII + 122 p., 34 fig., 5 tab., 1974.
ISBN 3-8055-1699-1
Vol. 4: GRIFFIN, C.J. and HARRIS, R. (Sydney): The Temporomandibular Joint Syndrome.
The Masticatory Apparatus of Man in Normal and Abnormal Function.
VI + 205 p., 96 fig., 8 tab., 1975.
ISBN 3-8055-2106-5
Vol. 5: GOODSON, J.M. and JOHANSEN, E. (Rochester, N.Y.): Analysis of Human
Mandibular Movement. VIII + 80 p., 50 fig., 20 tab., 1975
ISBN 3-8055-1416-6
Vol. 6: LILIENTHAL, B. (Halifax, N.S.): Phosphates and Dental Caries.
VIII + 108 p., 9 fig., 42 tab., 1977.
ISBN 3-8055-2677-6

Cataloging in Publication
Myers, Howard M.
Fluorides and dental fluorosis
Howard M. Myers. – Basel, New York, Karger, 1978.
(Monographs in oral science, vol. 7)
1. Fluoridation 2. Fluoride Poisoning 3. Mottled Enamel I. Title II. Series
Wl M0568E v.7/WU 270.3 M997f
ISBN 3-8055-1412-3

Contents

Contents

Preface

One of the justifications for the Monographs in Oral Science series is to permit periodic examinations of topics which have appeared in the literature over a span of years. The 'tying together' function can serve the scientific community in ways which are impossible to achieve in most reviews. The emphasis in this monograph is on critical evaluation of the pertinent literature rather than encyclopedic coverage of the field. It would be difficult if not impossible for this author to do both and the former seems to serve the greater need. What has been attempted is an effort to bring under a single cover the old and new concepts concerning the amount of fluoride ingested daily and the relationship that this amount has to the appearance of dental fluorosis. It is the hope of the author, only that the assumptions of the past have been subject to scrutiny and that none have been allowed the indulgence of being beyond consideration. Some of these assumptions follow:

(1) That drinking water is the major source of fluoride ingestion.
(2) That drinking water contributes 1 mg of fluoride to the individual per day.
(3) That intakes of less than 0.5 mg of fluoride are of no consequence.
(4) That substituting drinking water and other beverages can be done with impunity.
(5) That Ca deficiency predisposes to dental fluorosis.
(6) That blood fluoride is constant at low levels of fluoride intake.
(7) That organic fluoride does not contribute to the daily fluoride exposure.
(8) That a placental barrier exists for fluoride.
(9) That maintaining average fluoride intakes below a critical level will prevent dental fluorosis.
(10) That if a little fluoride is good, more is better.

The author wishes to express his thanks to Dr. HAROLD C. HODGE who was (and always will be) his mentor and to Dr. FRANK A. SMITH who contributed important components of the literature. Thanks are also extended to Ms. TERESA BOYLE who patiently perservered in the typing of this manuscript.

HOWARD M. MYERS
Philadelphia, Pa.

I. Fluoride Intake

Introduction

The well-established relationship between the fluoride concentration of drinking water and a lowered incidence of dental caries has given rise to considerable conjecture concerning the amount of fluoride required to obtain the protective effect of this element. Adding importance to this matter is the also well-established relationship between the fluoride concentration of water and the appearance of dental fluorosis. While levels of 2 ppm of F are associated only with the mildest form of dental fluorosis, a margin of safety of 2 can be regarded as impressively small. Knowledge of the amount of fluoride provided in the prevention of dental caries as compared with the amount capable of producing clinically significant dental fluorosis is obviously a goal of some importance. Claims that fluoride is increasing in the environment via the ambient air and fallout into vegetation must also be considered in the above light.

Fluoride intake by humans is a result of ingestion of food, water, and inhalation of fluoride-containing air. In addition, there is a potential for exposure to fluoride incorporated into medicinal preparations. These will be considered separately.

Fluoride Intake from Foods

Mostly on the basis of extensive analysis by McCLURE [1949] foods have been regarded as a small contributor to the total daily fluoride intake. McCLURE indicated that most foods are low in fluoride, having a content of 0.1–1.0 ppm F of dry weight, and would contribute a maximum of 0.27 mg of F/day. He was also among the first to observe that specific dietary items were high in fluoride and included in this list fish, tea and leafy greens.

Fish

Fish products generally have elevated fluoride levels when compared to meats. This is especially so if the fish preparation also includes the bones. Thus, mackerel, sardines and salmon may have fluoride levels close to 20 ppm on a dry weight basis. Shellfish in contrast are comparable to meats with levels of about 1 ppm. JONES [1971] analyzed the fluoride intake of a British population in West Midlands where fish represents a moderate portion of the diet, but includes the high fluoride items, mackerel and sardines. She found the minimum intake from fish to be 0.1 mg and the maximum 0.2 mg/day.

SAN FILIPPO and BATTISTONE [1971] from an analysis of an american diet (Baltimore), found the combined intake of fluoride from meat and fish to be 0.1–0.3 mg/day.

Tea

A second source of high fluoride is tea. The leaf itself is high in fluoride (about 100 ppm) but the infusions made from it are able to extract only a fraction of the total fluoride. Most investigators have given values of approximately 0.3 mg of F for an 8/oz serving of tea [HAM and SMITH, 1954; OELSCHLAGER, 1970; HARRISON, 1949; RAMSEY et al 1975]. The last author has made the interesting correlation that in 141 children of age 14 who consumed tea, the caries index (DMF) was in inverse relationship to the number of cups of tea consumed daily. MADDOCK [1969] has indicated that powdered instant tea can be somewhat higher than brewed tea, having almost 2× as much fluoride. He pointed out that the case of fluoride toxicity described by SAUERBRUNN et al. [1965], involved high consumption of instant tea, a quart of which contains 2.2 mg of F.

Leafy Greens

Plants, under ordinary conditions, contain 2–20 μg of F/g of dry wt [NAS, 1971]. An exception has already been mentioned, the family Theaceae, teas, camellias, etc., where the normal level is 50 to several hundred μg/g. Although soil is the chief normal source of plant fluoride, it is not possible to make direct predictions of plant fluoride content from soil levels alone. This is because acidic soils yield higher F contents in plants, and liming of such soils reduces the F content of the crop. In 4 different Utah soils the F levels in alfalfa were all 15–18 ppm (μg/g dry wt). Increasing the F levels of the soils to 200 ppm caused a doubling in the alfalfa grown in noncalcareous (acidic) soil but caused no effect in the 3 other soils. Even levels of 800 ppm in the calcareous soils caused only slight elevations of fluoride in alfalfa grown in them, while the same level considerably reduced the yield of alfalfa in the noncalcareous soil. Concentrations as high as 1,600 ppm were needed to increase the F levels and reduce the yield of the alfalfas grown in calcareous soils. There is about 8 times the average concentration found in US soils [WEINSTEIN, 1977]. The low sensitivity of plants to all but large changes of F in soils has suggested to most workers in this field that soil contamination can have only a minor effect on the F content of plants [NAS, 1971].

Plants themselves show highly variable sensitivity to fluoride. Visible changes in the leaves such as necrosis and chlorosis appear in the most sensitive group at levels of 20 ppm (gladiolus). Others, such as cotton, show no changes with levels as high as 4,000 ppm in the leaves. In general, the most sensitive plants show foliar response at levels of 50 ppm or less. Somewhat more resistant plants have a threshold of injury of 200 ppm or greater.

Plant F presents a problem for foraging animals – especially dairy cows. However, the established facts of storage of F in the skeleton and its low concentration in soft tissues and in milk makes this animal source of F a minor one for the human dietary [HODGE and SMITH, 1977].

Since the consumption of plants by humans is selective and since the fluoride content of most fruits and vegetables is low, about 1 ppm or less, this source is also of little concern, unless unusual dietary patterns prevail. Plants store fluoride in the margins of their leaves, so it might be expected that the leafy greens eaten by man are of some importance in this matter. Both spinach and cabbage are resistant to foliar signs of F content which means they will not exhibit necrosis until levels of fluoride of about 200 $\mu g/g$ dry wt are reached. It is therefore possible that such plants could be included in the human dietary without visible signs of damage. A 100-g portion of these fresh vegetables has about 90% water leaving a total dry wt of 10 g. The 10 g thus might have as a maximum 2,000 μg or 2 mg per serving. Aside from the improbability of such an occurrence, actual measurements on leafy greens indicate very much lower levels in the leaves. Both OELSCHLÄGER [1970] and McCLURE [1949] give values of 10 μg per serving.

JONES et al. [1971] has also provided an analysis of the F content of leafy greens such as cabbages, lettuce and brussels sprouts which were exposed to fluoride emissions in the midlands region of England. Her analyses showed a range of F content of 12–26 $\mu g/g$ dry wt.

Using a value of about 75–100 g of wet wt for daily consumption, the mg of F consumed in cabbage was 0.009 (average) and 0.023 (maximum).

For brussels sprouts these values were 0.021–0.053 and for lettuce 0.003–0.004 mg. She also showed that F levels for these vegetables only rarely exceeded 20 μg/g dry wt; higher values than this being characteristic only of unwashed samples.

Washing reduced the F levels by about one-third to one-half. Using maximum figures, JONES calculated that the total daily contribution for such contaminated greens would be 0.08 mg; while using average figures, this would be 0.03 mg for all 3 vegetables. Salad once a day in the form of 50 g of lettuce with an F content of 20 μg/g dry wt would contribute 0.05 mg daily.

OELSCHLAGER [1970] has also compared the effects of fallout of F wastes on vegetable products. Leafy products as expected had the highest levels, exceeding 20 μg/g. For lettuce the normal level found was 4.0 μg/g dry wt, while in samples exposed to F emissions the level was 77 μg/g. The so-called normal values are higher than reported elsewhere, even by the same author; 1 μg/g or less being the usual levels reported for lettuce. OELSCHLAGER also reported the presence of visible damage on the leaves of digitalis plants when the fluoride content was 31 μg/g; on pine needles when the F content was 37 μg/g; on sycamore, red beech, linden and chestnut leaves at levels of 200–400 μg/g. Other plants showed no visible damage at levels of 1,500 μg/g. These observations confirm the wide range of sensitivity found in the plant kingdom.

Fluoride in Plant Roots

As indicated above the chief source of fluoride from plants is by way of the storage of this element in the leaf margins. An exception to this rule has been reported for plants exposed to the preemergence herbicide trifluralin. This trifluoromethyl compound when added to the soil can be found as a residue in certain food plants such as onions, turnips and carrots. The amount found in carrots raised in such soil was 0.25 ppm in the tops and 0.65 ppm in the roots. Trifluralin and its n-propyldealkylated metabolite, both of which are fully fluorinated, represent the major products found. 70% of the amount in the roots was located in the outermost $1/16$-in layer [GOLAB et al., 1967]. Trifluralin is metabolized by animals without any conversion to inorganic fluoride from the trifluoromethyl group [EMMERSON and ANDERSON, 1966]. No expired CO_2 from the ^{14}C-labelled CF_3 group was found, and 100% of the label appeared in urinary plus fecal products. The plant roots themselves are able to convert a small amount, about 5%, of the organic fluorine to inorganic fluoride as shown by recovery of a defluorinated metabolite and by their ability to produce labelled $^{14}CO_2$ from the $^{14}CF_3$-labelled trifluralin molecule. The possibility that food crops might contain inorganic fluoride from this herbicide is clearly possible but the amount would be quite small. A 100-g serving of carrots with 0.65 ppm of trifluralin would contain about 5% of the total trifluralin as F or 0.003 mg. In the leafy or top portions defluorination occurs to about 50%, thus a serving of 100-g of turnip greens might contain 0.001 mg.

Effect of Cooking

One additional source of food fluorides is not related to atmospheric contamination but is pertinent to the discussion. It is the effect of cooking foods in 1 ppm F containing water. OELSCHLAGER [1970] has reported that certain foods take up significant amounts of F on being cooked in such water. Polished rice had 10 μg/g (1 mg/serving) of F compared to 0.24 μg (24 μg/serving) in the raw state; polished peas likewise had levels of 1.5 mg/serving after cooking compared to 12 μg/serving in the raw state. These figures are in the range reported for leafy green vegetables (10–20 μg/g) and could represent a similar hazard if eaten to excess. Although cooked potato and cauliflower are said to behave similarly, numerical data were not given for these items.

MCCLURE [1943] calculated the amount of fluoride that fluoridated water would contribute to food. For varying age groups he estimated from 0.1 to 0.4 mg/day would be added by preparing food in fluoridated water.

MARIER and ROSE [1966] measured the increase in food items prepared in fluoridated water and found on the average, an increase of 0.5 ppm. On the basis of a single serving being 100 g this converts to 0.05 mg (50 μg) per serving. The data provided includes potato, green beans, mixed vegetables, kernel corn, carrots and green peas and tomato soup. With the obvious exception of the last item, these increases were in the solid portions. Analyses for the liquid and solid portions gave good evidence that an equilibrium had been approximated between the two.

The role of the cooking vessel composition on the fluoride content of water was studied by FULL and PARKINS [1975]. When the volume of water was reduced by 50% by boiling for 15 min, the fluoride content of water in a Teflon pan was increased to about 3 ppm from the original 1 ppm. This suggests some release of F from the vessel or at least, no loss of fluoride by reaction with the vessel. Aluminium caused such a decrease in fluoride after similar boiling, levels of 0.3 ppm being measured in the remaining $^2/_3$ of the water. Stainless steel and Pyrex produced no change other than that expected by the concentration due to boiling.

Fluoride Contribution of the Diet

The contribution of foods to the daily intake of fluoride has been directly measured by several authors. These findings are based on representative diets such as are available in metabolic wards or as over-the-counter purchases.

MACHLE *et al.* [1942] reported the average fluoride intake from food for one individual over a 4-month period to be 0.16 mg/day. For fluids it was 0.3 giving a total of 0.46 mg/day. The highest food-borne intake over a 24-hour period was 0.5 mg while for fluids it was 0.8. The diet was one chosen by the subject which avoided fish, fish products and salted foods. The water supply was unfluoridated with a fluoride level of less than 0.1 ppm.

HAM and SMITH [1954] analyzed the diets of 3 young women aged 23–24 years over a 3-day period. The diets prepared with unfluoridated water were considered normal but did not include fluoride-rich items such as fish or tea. On 3 successive days the total fluoride contributed by these meals was 0.43, 0.76, and 0.45 mg.

SAN FILIPPO and BATTISTONE [1971] analyzed 4 different dietaries used by young adults aged 16–19 years. They found fluoride intakes from foodstuff only, prepared in fluoridated water to be 0.88, 0.90, 0.78 and 0.83 mg/day.

Table I. Analyzed fluoride intake from food sources in fluoridated and nonfluoridated areas

	Minimum	Maximum	References
Nonfluoridated			
A	0.16	0.50	MACKLE [1942]
B	0.43	0.76	HAM and SMITH [1954]
C	0.78	1.03	KRAMER *et al.* [1974]
D	0.91	1.5	LEE [1975]
Average	0.67		
Fluoridated			
E	0.78	0.90	SAN FILIPPO and BATTISTONE [1971]
F	1.7	3.4	KRAMER *et al.* [1974]
Average	1.24	2.15	
Difference	0.67	1.20	

Diets A and B avoided fish and fish products. Diet D was high in fish products.

KRAMER *et al.* [1974] found the dietary fluoride from foodstuff only in 16 different hospitals around the US to range from a low of 1.7 to a high of 3.4 mg in areas where water was fluoridated. The range of values for 4 non-fluoridated cities was 0.78–1.03 mg.

Table I which summarizes these data is restricted to only those publications in which actual analysis of dietary fluoride was performed. The publications by MCCLURE [1943] and MARIER and ROSE [1966] in which estimates are given are not tabulated. Table I indicates that food-borne fluoride is as variable as might be expected. Two factors are singled out in these data, the contribution of fluoride-rich foods such as fish, and the contribution of fluoridated water to food fluoride. Comparison of diets B and D indicates that a diet rich in seafood can contribute between 0.50 and 0.75 mg of fluoride/day. The contribution of food to the daily intake of fluoride in the absence of any additional fluoride in the drinking water is between 0.6 and 1.0 mg day. Fluoridated drinking water can be expected to add to this between 0.6 and 1.2 mg of F. It will be seen subsequently that up to 1 mg more per day can be expected from all sources of water-borne fluoride.

Fluoride Intake from Fluids

Water-borne fluoride has been said to represent the largest single component of this element's daily intake. To assess the contribution of water fluoridation to daily fluoride intake, however, it is necessary to separate total daily fluids into several components: that portion consumed directly as drinking water, tap water used in preparation of other foods and non-tap water fluids. The three components contain differing amounts of fluoride and these amounts can be expected to vary with age.

MCCLURE [1943] approximated answers to the above question by using the physiological generalization that water requirement in milliliters was equal to the caloric requirement per day. He developed a table in which the drinking water consumption was assumed to range between 25 and 33% of the total water requirement. He then added components from water used in preparation of food and from food. Assembling the age range 1–12 into four separate groups, he calculated the range of drinking water intake to be between 500 and 1,100 ml/day.

More direct determinations of water use have been made with the use of questionnaires [MCPHAIL and ZACHERL, 1965]. His findings have been converted into milliliter units and are presented in table II. This table has

Table II.

Age, years	Drinking H_2O	Infant formula	Reconstituted juice	Soup	Tea	Other	Carbonated beverage	Whole juie	Milk	Reconstituted juice	miscellaneous	Total, ml
a Water consumption (in ml) per person per day, by age [McPHAIL and ZACHERL 1965]												
0.5–1	35	146	20	1	–	5	–	64	557	118	18	964
1–2	115		33	22	3	24	13	59	355	81	23	728
3–4	197		55	29	7	31	16	41	376	114	7	873
5–6	238		60	32	14	12	13	48	332	173	13	935
7–8	314		59	41	8	17	48	38	339	193	12	1,069
9–10	510		74	57	8	27	37	51	368	167	11	1,310
b Fluoride consumed (in mg) in each of 11 categories of fluid consumption per person per day, by age												
0.5–1.0	0.035	0.102	0.02	0.00	–	0.005	–	0.079	0.111	0.142	0.018	0.452
1–2	0.115		0.03	0.02	0.003	0.02	0.001	0.018	0.071	0.097	0.023	0.578
3–4	0.197		0.06	0.03	0.008	0.03	0.001	0.012	0.075	0.137	0.007	0.557
5–6	0.238		0.06	0.03	0.017	0.01	0.001	0.014	0·066	0.208	0.013	0.657
7–8	0.314		0.06	0.04	0.01	0.02	0.034	0.011	0.068	0.232	0.012	0.801
9–10	0.510		0.07	0.06	0.01	0.03	0.026	0.015	0.074	0.200	0.011	1.006

converted fluid ounces per person per day into milliliters per person per day. It may be observed in these data that total fluid intake from all tap water sources and other sources ranges from 700 to 1,300 over the age span 1–10 years. Of this amount, milk from all sources represents at least one-half of the total fluid at all age groups. At age 9–10 drinking water and milk are each about 50% of the total fluid intake, but at younger ages milk represents more than 50% of the total fluid ingested.

Table II has been supplemented with additional calculations to determine the amount of fluoride ingested daily from the eleven categories of fluid listed. To do this the level of fluoride found in each category has been obtained from the literature or if not available has been assumed to be 1 ppm. With the assumption that 1 ppm of fluoride is present in the water supply, the F concentrations used for computation are as follows: drinking water, 1 ppm; formula, 0.7 ppm [WIATROWSKI et al., 1975]; reconstituted juice, 1 ppm [WIATROWSKI et al., 1975]; soups, 1 ppm; tea, 1.2 ppm [HARRISON, 1949]; carbonated beverages, 0.7 ppm [MARIER and ROSE, 1966]; whole juice, 0.3 ppm [OELSCHLAGER, 1970]; whole milk, 0.2 ppm [ERICSSON and RIBELIUS, 1971]; reconstituted milk, 1.2 ppm [WIATROWSKI et al., 1975].

Table IIb reveals that the fluoride contributed from all sources of fluid in the age range of 1–10 years varies directly with age from 0.450 to 1.0 mg/person/day. It is of some importance to note that in this age range, drinking water does not add more than 0.5 mg of fluoride/day. Total fluoride from all liquids does provide as much as 1 mg/day in the 10-year-old group, but drinking water *per se* does not.

The single most important factor in determining water consumption is the maximum daily temperature in the area consuming the water [GALAGAN and LAMSON, 1953]. Table II is derived from a single community, Edmonton Canada, and cannot be used without further consideration of climate. This has given rise to table IIIa, which lists the drinking water consumption and the total fluid consumption by age, in five different cities located on a north to south axis. These data taken from WALKER et al. [1963] and McPHAIL and ZACHERL [1965] indicate that drinking water reaches the level of 1 liter/day only in one instance; and that drinking water represents about 50% of the total fluid intake in the older age groups. In accord with table IIa, total fluid intake is approximately 1 liter/day.

Table IIIb presents the average, for all age groups, of drinking water and total fluid intakes on a geographic basis. No clear cut trend from north to south is evident.

In table IIIc, the average drinking water and total fluid intakes are listed

Table IIIa. The effect of geographic location (climate) on drinking water and total fluid consumption

Age, years	Location	Drinking water, ml/day	Total fluids, ml/day
0–1	Edmonton	35	964
	Kalamazoo	79	861
	Los Alamos	328	699
	Atlanta	259	743
	Miami	270	725
1–2	Edmonton	115	728
	Kalamazoo	283	832
	Los Alamos	304	751
	Atlanta	229	743
	Miami	216	789
2–3	Edmonton	–	–
	Kalamazoo	305	894
	Los Alamos	284	838
	Atlanta	384	977
	Miami	445	1,038
3–4	Edmonton	197	873
3–5	Kalamazoo	305	889
	Los Alamos	394	960
	Atlanta	396	1,007
	Miami	284	845
5–6	Edmonton	238	935
5–8	Kalamazoo	296	1,051
	Los Alamos	466	1,088
	Atlanta	464	1,247
	Miami	390	1,109
7–8	Edmonton	314	1,069
9–10	Edmonton	510	1,310
8–12	Kalamazoo	–	–
8–12	Los Alamos	551	1,347
8–12	Atlanta	1,008	1,660
8–12	Miami	265	1,144

by age. A definite trend toward increasing consumption of both drinking water and total fluid is evident with increasing age from 1 to 12 years.

Milk is known to have a much lower fluoride concentration than water whether or not fluoridation at the level of 1 ppm exists. Milk F levels are usually given as 0.07–0.22 ppm [McClure, 1949]. These data have been included in table II. A special situation exists in for infants who are bottle fed.

Table IIIb. Average for all ages (1–12) of drinking water and total fluid consumption on the basis of north to south axis

Location	Drinking water, mɪ/day	Total fluids, ml/day
Edmonton	521	979
Kalamazoo	326	905
Los Alamos	388	947
Atlanta	457	1,063
Miami	312	942

Table IIIc. Average, for all locations in five cities, of drinking water and total fluid consumption on the basis of age

Age, years	Drinking water, ml/day	Total fluids, ml/day
0.5–1	241	798
1–2	266	768
2–3	355	937
3–5	364	915
5–8	467	1,083
8–12	670	1,365

While milk is low in fluoride, formulas made with evaporated milk and tap water containing 1 ppm fluoride have significantly high levels of F⁻. Formula made from fluoridated water ranges in fluoride concentration from 0.4–0.9 ppm. This may be reduced to lower levels by the use of distilled water in reconstituting the formula [WIATROWSKI *et al.*, 1975]. Infants less than 6 months of age have total dietary fluoride intakes ranging from 0.32 mg/day at 1–4 weeks of age to 1.23 mg/day at age 4–6 months, if fed on reconstituted juice and milk formulas. ERICSSON and RIBELIUS [1971] have pointed out that cows milk is higher in fluoride than human milk, 0.2 vs. 0.025 ppm, and that when 1 ppm water is used for dilution the preparation contains 20 times more fluoride than human milk. Dry formulas have even higher levels of fluoride ranging between 1.07 and 1.33 ppm. When these are used in formulas the fluoride levels are about 50 times that of breast milk. These calculations have not been included in table II since they apply to ages up to 6 months only, represent a rather limited period of exposure, and are highly individual.

Fresh fruit juices are also low in fluoride content, the range being 0.1–0.3 ppm. Reconstituted fruit juices, however, made with fluoridated water vary in fluoride concentration from 0.3 to 2.5 ppm [WIATROWSKI *et al.*, 1975]. The calculations for reconstituted fruit juices in table II therefore used a basis of 1 ppm for fluoride content. Bottled beverages only partially reflect the concentration of fluoride in the water where they are prepared. MARIER and ROSE [1966] found that an 8 oz serving of ginger ale made with 1 ppm fluoridated water raised its fluoride content from 5 μg (unfluoridated) to 190 μg (0.7 ppm). Table II utilizes a value of 0.7 ppm.

TOTH [1975] has calculated the amount of dietary fluoride needed to substitute for the total fluoride contributed by drinking water. His calculations are based on body size, caloric intake, fluid intake and fluoride derived from food. Data are provided for optimal dose, tolerable dose and toxic dose for 3 age groups. His calculations have been converted to total fluoride by multiplying the weight in kilograms for each age group by the mg/kg allowance.

For infants less than 1 year of age with a weight of 10 kg, the optimal dose in mg of F is 0.45; for children age 7–9 (24 kg) the dose is 0.77; for adults (65 kg) the dose is 1.43 mg. It will be seen that these figures are compatible with those of table IIb dealing with water-borne F for children.

Excessive consumption of drinking water might be expected to increase fluoride intake under special circumstances. Such findings have been reported for Lucknow India where children have been shown to utilize drinking water for as much as 85–90% of total fluid intake. This study is discussed in chapter II.

An opportunity to study the effect of increased water consumption with fluoridation at the 1 ppm level over a 6-year period was found by POT and FLISSEBAALJE [1974] in a glass foundry in the Netherlands. Office workers were found to consume from 2.6 to 3.0 liters/day while workers more directly exposed to the heat of the foundry consumed water at the level of 4.0–6.0 liters/day. Fluoride intakes were determined and found to be 2.0–2.7 mg F/day for the office workers and 3.6–4.7 mg/day for the heat-exposed workers. The plasma fluoride of the office workers was 1.58 μM/l while the heat exposed group had an average level of 4.47 μM/l. The plasma data indicate that the somewhat elevated water consumption of the office workers did not cause an elevation of the plasma fluoride over that of individuals consuming lower volumes of 1 ppm water. The heat-exposed workers did show excess intake of fluoride via drinking water and this was confirmed by the urine analyses. These indicated that both groups were receiving fluoride at higher than previous levels since F excretion was 50% or less. The 2.0–2.7 mg of F/day obtained from the drinking water by office workers was therefore

Table IV. Total daily fluoride intake reported by various authors

mg F/day	Conditions	Author
0.10–0.17	formula diet 6 to 14 week-infants	HAM and SMITH [1954]
0.06–0.19	breast fed infants	ERICSSON [1972]
0.46	normal diet, young men, balance study	MACKLE [1942]
0.42–0.82	estimates based on caloric requirements	McCLURE [1943]
0.43–0.79	selective diet, avoided tea and fish	HAM and SMITH [1954]
0.65	nonfluoridated area, young individuals	AUERMANN and BORIS [1971]
0.75	army field rations	SAN FILIPPO and BATTISTONE [1972]
1.2	British children	MURRAY et al. [1956]
1.2	nonfluoridated area, hospital meals	KRAMER et al. [1974]
1.3	nonfluoridated area, adults	AUERMANN and BORIS [1971]
1.2–1.4	test diet with 1,500 ml of tea	HAM and SMITH [1954]
1.5	normal diet, adults (4) balance study	McCLURE et al. [1945]
1.4–1.9	formula fed infants using 1 ppm H_2O	ERICSSON [1972]
1.72	children in Lucknow India fluorosis area	NANDA et al. [1974]
1.99	Guernsey children 7–16 years, nonfluoride	COOK and FRANCE [1976]
2.0–2.7	office workers in glass foundry consuming 3 liters of 1 ppm H_2O	POT and FLISSEBAALJE [1974]
2.2	British women	MURRAY et al. [1956]
2.2	16–19 year olds in Baltimore, ppm fluoride	SAN FILIPPO and BATTISTONE [1971]
2.24	British children 5–15 years, nonfluoride	COOK and FRANCE [1976]
2.3	fluoridated area, young individuals	AUERMANN and BORIS [1971]
2.7	Newfoundland regular diet with tea	ELLIOTT and SMITH [1960]
2.9	fluoridated area, adults	AUERMANN and BORIS [1971]
3.2	British men	MURRAY [1956]
3.6	fluoridated areas, hospital meals	KRAMER et al. [1974]
3.8	basal German diet	OELSCHLAGER [1970]
3.0–4.0	Australian children with water at 1.5 ppm	KALLIS and SILVA [1970]
1.9–4.0	fluid intake 1–3 liters, food 1 mg	MARIER and ROSE [1966]
2.9–5.0	fluid intake 1–3 liters, food 2 mg	MARIER and ROSE [1966]
5.0	fluids 0.5 mg/day, food estimated at 10 × that, Japan	OKAMURA and MATSUSHIDA [1969]
3.57–5.37	basal diet without fish or tea in a balance study	SPENCER et al. [1970]
3.6–4.7	glass factory foundry workers consuming 4–6 liters of 1 ppm H_2O	POT and FLISSEBAALJE [1974]

contributing more fluoride daily than could be obtained with drinking water volumes of approximately 1 liter. While this excess failed to be detected in blood levels it was reflected in the renal excretion of these individuals. We may conclude that increased water consumption when 1 ppm fluoridation is present will add more fluoride to the skeletons of those consuming it but blood levels will not rise unless the volume of water consumed exceeds 3 liters/day for a period of several years.

Table IV provides a summary of the total fluoride intakes for humans from water and food-borne sources as reported in the literature since 1943. The higher values of 4–5 mg at the end of the table are subject to some doubt. As will be seen elsewhere in the text, they represent either special circumstances of unusual consumption or assumptions not otherwise supported.

Airborne Fluoride

Airborne fluoride occurs as dusts, as gaseous fluorine, and HF. Analyses of 7,700 air samples from localities throughout the United States for total water-soluble fluorides were made in 1966 and 1967 by the National Air Pollution Control Administration. 97% of the samples from nonurban areas had no detectable fluoride. The highest concentration found in these samples was 0.16 μg/m^3 [NAS 1971].

In urban areas 87% had less than 0.05 μg/m^3 (the lower limit of detection). The ranges found were from 0.05 to 1.89 μg/m^3.

Since an individual consumes about 20 m^3 of air per day, the dose of F from this source can be calculated. The highest level found (1.89 μg/m^3) will yield 0.0378 mg/day. The highest level found in nonurban air would yield 0.0032 mg/day. Thus, the total contribution per day of airborne F is less than 0.04 mg at the maximum air concentration found. In certain selected areas where F from industry is heaviest, the above values are exceeded. Measurements made in 1967 in the immediate vicinity of an aluminium factory near The Dalles, Oregon gave levels of up to 5 μg/m^3. This is equivalent to 0.1 mg/day if this air were inhaled steadily over that period. Occasional single determinations of 2–3 × the above level have been reported around industrial areas but the level usually found is 5 μg/m^3 or less. The most extreme levels found were those of the aluminium factory of Fort William, Scotland during WW II when scanty consideration was given to worker safety. The furnace room concentration was 1.42 mg/m^3, 2,000 yards away the air sample was 0.22 mg/m^3 and a mile away it was 0.042 mg/m^3

[NAS, 1971]. Where exceptions such as this exist [HODGE and SMITH, 1977], the exposure involves residence within a mile downwind from aluminium or phosphate production emissions, or obviously, workers directly exposed to such emissions.

Pharmaceutical Preparations

Pharmaceutical preparations may introduce fluoride either deliberately or inadvertently. In the former case are included sodium fluoride tablets, dentifrices or mouthwashes which contain fluoride. The first of these is used as a substitute for water fluoridation and is compounded to provide up to 1 mg of fluoride ion per day on the basis that this amount is provided by 1 liter of 1 ppm water. The obvious difference between 1 mg administered over 24 h in drinking water and 1 mg taken all at once in tablet form has raised some doubts as to the equivalence of the two preventive regimens. While it is reasonable to assume that a 1-mg tablet therapy produces a definite peak value in blood, this is more easily demonstrated for doses greater than 1 mg of fluoride [EKSTRAND, 1977a]. In the absence of administered fluoride, blood levels of 0.01–0.02 ppm (0.5–1.0 $\mu M/l$) are the rule. To elevate this baseline level with a dose of 1 mg of fluoride requires the initiation of a standardized diet of low fluoride and limited fluid 24 h before and during the test period. A peak of 0.05 ppm (2.60 $\mu M/l$) will occur at 30 min [HENSCH-LER, 1975]. The difference anticipated therefore between aqueous and tablet regimens can be demonstrated. Caries prevention with different regimens have not always been equivalent [WIRZ, 1964]. Patient compliance is markedly less with tablets than with water fluoridation [FANNING et al., 1975].

Other effects, such as rate of excretion, have shown some differences in young children which have been attributed to the vitamin component rather than to the tablet form. MARGOLIS et al. [1967] compared dissolved fluoride, tablets of NaF and the combination of fluoride and vitamins in children aged 1–4 years who had no prior exposure to fluoride. Over a 3-day period, 1 mg of fluoride was excreted to the level of 50% when dissolved F or plain NaF tablets were used. A slight delay in the elimination of fluoride was found for the dissolved preparation but at the end of the 3-day period no difference in excretion remained. The vitamin supplemented fluoride, in contrast, showed less excretion than the unsupplemented preparations: only 30% rather than 50% of the ingested dose appeared in the urine at 48 or 72 h. These data can be interpreted to suggest that retention of fluoride was greater in these young children when vitamin supplements were included with the fluoride, since a 3-day period is usually sufficient to recover un-retained fluoride. The additional amount of fluoride so retained is about 0.3 mg. Long-term studies over 3 years confirmed the additional retention of F in young children given vitamin-supplemented fluoride [STOOKEY, 1970], 35% of the dose being excreted in the urine.

An effect of a vitamin supplement on fluoride retention in young adults could not be demonstrated by HENNON et al. [1969]. Retention of 45–55% was found for both vitamin-supplemented and unsupplemented NaF tablets in these individuals with no prior fluoride exposure. The differences observed in the 2 age groups may be due to the greater retention of fluoride in younger individuals reported by EKSTRAND [1977a]. He attributed this to the greater degree of skeletal saturation by F in older individuals.

Dentifrices and Mouthwashes

Dentifrices and mouthwashes, unlike the tablet preparations, are not intended to replace water-borne fluoride and therefore are not intended to provide additional systemic fluoride. The daily use of such fluoride-containing dentifrices has been promoted by the dental profession as a means of preventing dental caries. Although the net reduction in the incidence of dental caries has not been uniformly impressive with the use of these agents, their use is growing, aided by joint efforts of the manufacturers and practitioners. Although fluorides are present in high concentration in these dentifrices, the intended use of these agents is to allow only for a topical effect on the teeth rather than a systemic one. Since an important group of recipients of these preparations is children, it is appropriate that the ingestion of fluoride from dentifrices be shown not to be excessive for this group especially. The question of the fluoride contribution of such dentifrices to the total daily intake of this element is an important one since the preventive programs rely on regular daily use of these agents.

Dentifrices contain in the fresh product, approximately 1,000 ppm of fluoride, the range being 900–1,100 ppm. The presence of reacting substances causes some loss of fluoride ion over a period of time of approximately 4 months, after which the level of fluoride ion becomes relatively stabilized at a lower level. Since a soluble fluoride level of 150 ppm has been shown to be effective in clinical trials, levels above this are regarded as evidence of the effectiveness of the dentifrice even after aging. The remaining ionic fluoride at the recommended expiration date is greater than 250 ppm for any of the products on the market and may range from 250 to 900 over a period of from 2 to 5 years.

Any assessment of the daily fluoride contribution of dentifrices must concern the amount of dentifrice ingested with each use. BARNHART et al. [1974] have made a study of 4 different age groups of 50–75 patients each, utilizing a total recovery method of assessing ingestion. This method tends to err on the high side since any failure to recover used or expectorated dentifrice will be considered as ingestion. These investigators also used duplicate and triplicate brushings by the same individuals to assess within patient as well as between patient variability. Their findings were reported as the grams of dentifrice ingested for 3 age groups along with predictions of the 90th and 95th percentiles for the amount of dentifrice ingested. The amount of dentifrice applied to the brush was close to the same value for all groups, about 1 g. The amount ingested varied inversely as the age of the subjects. Table V indicates the findings of BARNHART et al. along with calculations of the amount of fluoride that would have been ingested with the dentifrice if it had been present. In the youngest group (5–7) for which complete analysis is provided, two brushings a day would contribute a maximum of about 0.5 mg of fluoride. ERICSSON and FORSMAN [1969] used 0.5 g of two different fluoride-containing dentifrices on children aged 4–5 and 6–7. The younger group retained from 26 to 33% while the older group retained 25–28% of the fluoride. In milligram amounts, the retentions were 0.13 and 0.12 mg per use. On a twice a day basis this would be 0.25 mg/day. Since 0.5 g of dentifrice was used in comparison to the 1.0 g in other studies, the findings as to fluoride intake from dentifrices appear to be in reasonable agreement.

All the fluoride ingested cannot be assumed to be absorbed. The presence of considerable excess Ca and PO_4 in dentifrices might have a retarding effect on the absorption of the ingested fluoride. Animal studies by DESHPANDE and BESTER [1964] have indicated an absorption of only about 30% of the fluoride ingested with dentifrices. In human adults, absorption has ranged from means of 3 to 12% (maximum 35%) using radiofluorine-labelled dentifrices [ERICSSON, 1961]. In children, HARGREAVES et al. [1970] found that for a single dose of fluorophosphate, the urinary excretion was 20%. This would suggest an absorption of 40% according to the widely accepted generalization that urinary fluoride represents about ½ of the absorbed amount of fluoride in subjects who are not in a steady state.

Of singular importance has been the question of the amount of fluoride ingested by children of age group 3–6. HARGREAVES et al. [1970, 1972] studied

Table V. Medicinal sources of fluoride

Vehicle	Age, years	Total mgF exposure	Amount	Retained mgF per use	Reference
Dentifrice	2–4	860	1 g of paste	0.82	BARNHART et al. [1974]
	4–5	500	0.5 g of paste	0.13	ERICSSON and FORSMAN [1969]
	5–7	940	1.0 g of paste	0.24	BARNHART et al. [1974]
	6–7	500	0.5 g of paste	0.12	ERICSSON and FORSMAN [1969]
	11–13	1,110	1.0 g of paste	0.13	BARNHART et al. [1974]
	20–35	1,139	1.0 g of paste	0.07	BARNHART et al. [1974]
Mouthrinse		Volume, ml	Time, min		
	5–7	50	1.0	3.0–5.0	HELLSTROM [1960]
	5–7	10	1.0	0.85	HELLSTROM [1960]
	5–7	3.5	0.5	0.0	BIRKELAND and LØKKEN [1972]
	4–6	1.8	0.5	0.15	FORSMAN [1976]
	4–6	1.8	1.0	0.23	FORSMAN [1976]
	3–5	1.5	0.5	0.3–0.4	ERICSSON and FORSMAN [1969]
	5–7	1.5	1.0	0.3–0.4	ERICSSON and FORSMAN [1969]
	adults	75	1.0	1.5–5.4	HELLSTROM [1960]
	adults	15	1.0	2.0	HELLSTROM [1960]
F/vitamins			Form		
	1–4	1.0	tablets – 3 days	0.7	MARGOLIS [1967a]
	young adults	1.0	tablets	0.5	STOOKEY [1970]
General anesthesia			Agent		
	adults	5 hrs.	methoxyflurane	3.0	DOBKIN and LEVY [1972]
	adults	5 hrs.	halothane	0.06	CREASSER and STOELING [1973]
	adults	5 hrs.	fluoroxene	0.06	CREASSER and STOELING [1973]

dentifrice absorption by using two different methods of assessing it. In one study, urinary excretion was used to determine the amount of dentifrice ingested. Using a test dose of 2 mg of fluoride they determined that 20% of the amount ingested was excreted in the urine. They concluded that the 90th percentile ingested 0.50 g of dentifrice per day: at fluoride concentrations of 1,000 ppm this would yield F intakes of close to 0.5 mg/day, a figure which while not included is in accord with table V. These values, however, are confused by the fact that variable dietary fluoride was being consumed during the time the study was performed. This resulted in unpredictably high urinary fluorides in some subjects and required the use of an estimate to allow for the dietary contribution.

Using a gravimetric method that attempted full recovery of the polishing agent, a similar study was performed on the same age group [HARGREAVES et al., 1972]. This method tends to overestimate ingestion since any unrecovered paste is included in that category. The data reveal that 0.5 g/day was ingested by the 70th percentile. Here too, corresponding fluoride intakes from fluoride containing dentifrices in young children could be as high as 0.5 mg/day.

Mouthwash preparations containing fluoride are also popular components of preventive programs. These formulations contain no abrasive Ca salts and are therefore likely to result in more complete absorption of any ingested fluoride. HELLSTROM [1960] found that with a concentration of 0.5% of NaF, the halogen was excessively retained by both adults and children. Of the 50 mg of fluoride used per rinse, 3–5 mg could not be accounted for in children 5–7 years of age. For adults as much as 6 mg of 75 used per rinse could not be recovered. HELLSTROM concluded that a lower concentration was indicated and selected a 0.1% solution. Unrecovered fluoride in this latter study was 0.85 mg of the 10 mg used for children and 2 mg of the 15 used for adults. These amounts are closer to the level of daily ingestion of fluoride found when drinking 1 ppm fluoridated water. HELLSTROM's data indicate that in a few cases as much as 1.5–2.0 ml of the rinse itself was not recoverable and it was these same individuals who had the highest unrecovered fluoride. BIRKELAND and LÖKKEN [1972] lowered the amount of fluoride used per rinse to 3.5 mg by reducing the concentration to 0.05%, volume to 7 ml, and the rinsing time to 30 sec. They were able to identify these conditions as the ones under which no detectable retention of fluoride occurred. TORELL and ERICSSON [1965] have reported that the 0.05% NaF mouthwash used above is an effective anticaries agent.

ERICSSON and FORSMAN [1969] also assessed the potential intake of fluoride in young subjects. In the age range below 3, the swallowing reflex is insufficiently developed to allow even 7 ml of mouthrinse to be recovered satisfactorily. Of 7 ml of water used only 2 ml were recovered in the expectorant after ½ min. In the age range 3–5, 5 of the 7 ml of rinse were recovered, while at ages 5–7 the recoverable amount was 6.5 ml. With a test rinse (0.054%) the retained fluoride was 0.3–0.4 mg (21–26%) of the 1.5 mg total. 7 ml of rinse were used for ½ min in the 3- to 5-year-old group and for 1 min in the 5- to 7-year-old group.

More recently, a mouthrinse containing 0.025% NaF was tested on 135 children in the age range 12–13 for a 2-year period. The caries rate was similar for a 0.025 and a 0.2% rinse. Retention of fluoride in a 4- to 6-year-old group of children using the lower concentration was 0.23 mg when a 1-min rinse was used, and 0.15 mg when the rinse was completed in ½ min [FORSMAN, 1974a].

It is concluded that mouthrinse may be so controlled as to contribute minimal amounts of fluoride. This can be done by reducing the concentration of F, restricting the volume provided per rinse and confining the rinse time to 0.5 min. Table V summarizes these data.

It would appear desirable to reduce the unneeded daily intake of fluoride to as low levels as possible. Given several preventive programs of approximately equal efficacy in preventing caries, those requiring the least ingestion and retention of fluoride should be preferred. Mouthrinses which can be defined as to concentration, duration and volume would appear to offer a greater control over exposure to fluoride than dentifrices applied to the teeth with a brush.

Organic Molecules as a Source of Fluoridation

Growing experience with syntheses involving C-F bonds generally gives support to the idea that they are highly stable metabolically and that they confer increased intensity of action to drugs in which they are present. The greater potency and toxicity reported for fluorine-containing molecules is due to the properties of the intact electronegative atom itself rather than to any release of fluoride ion. The conversion of fluorine-containing molecules to fluorine containing derivatives has been well documented [SMITH, 1970]. Fluorine-substituted groups have been shown to remain intact through several enzymatic transformations: the covalently bonded fluorine subsequently appearing in the principle metabolites.

While the statements that the C-F bond is a stable one is undoubtedly true, it does not follow that it is resistant to all possible metabolic reactions. It has been demonstrated unequivocally that inorganic fluoride can be recovered in mammals following the administration of some covalently bonded F-containing molecules. The amount of such cleavage is generally quite small, however – about 1% of the administered dose. Any fluoride intake hazard from this source, therefore, would depend on the nature of the drug, its dose and frequency of administration.

The discussion which follows will consider those agents which are currently in use and which might contribute daily fluoride increments to the individual. Strictly experimental fluorine-containing compounds, however, will be discussed first to provide the necessary generalizations needed to assess those therapeutic agents which have fluorine atoms.

Methods of Study

There are two basic methods for studying the conversion of organic fluoride to inorganic during metabolism. The first of these is classical and consists of labelling of the molecule in question with a radioatom and following the fate of the label. ^{14}C labelling appears to be the most valid and useful radionuclide for this purpose, the 2-hour ^{18}F half-life being too short for extensive metabolic studies. The form in which the ^{14}C label appears determines the fate of the C-F bond which was previously labelled. $^{14}CO_2$ derived from a ^{14}C-F bond, is unequivocal evidence of the release of fluoride by metabolic means. Secondary to this is the identification of fluorine containing metabolites in blood or urine. If these have been recovered after sufficient time for metabolic reactions to have occurred, they may be regarded as convincing evidence of stability of the C-F bond. It need not be emphasized here that the demonstration of one or more fluorine containing metabolites cannot rule out the possibility of subsequent or concurrent loss of the F atom. Thus only a complete quantitative metabolic scheme, a recognized rarity, can prove whether or not fluoride release occurs in those instances where F-containing metabolites are identified.

A second type of study dealing with fluoride release in test animals involves either the well-known affinity of fluoride for skeletal tissue or its effect on skeletal structures. A release of fluoride to the blood may be inferred from analysis of bone mineral which has acquired it as a result of growth or exchange. The bone level is directly related to present and to some extent, prior blood concentrations.

Alternatively, fluoride can be shown to be present by its effect on dental caries in test animals. For these studies, young weanlings are necessary as is a cariogenic diet. The latter is often nutritionally inadequate by virtue of its high sucrose or glucose content – about 60%. Although fluoride has a specific and measurable effect on dental caries other factors can be present which will complicate the interpretation of data. Thus, age of the animals, effect of the agent on salivary secretion, eating habits or alterations of them due to the medication, etc., all can modify caries data, and these may require the use of several control groups of animals.

Somewhat higher levels of fluoride can be detected by the specific bleaching effect of dental fluorosis in the growing incisors of rodents. In this test the usual orange-brown pigment seen on the labial surface of the incisors is bleached in either a striated pattern at low doses, or completely at higher doses.

Defluorination of Organic Molecules

Aliphatic Molecules

Monofluoroacetate. This compound has an established toxicity, convulsions being produced in 30 min by 5 mg/kg [GAL *et al.*, 1961]. The toxicity depends on the stability of its C-F bond. Its effects are considered as derived from inhibition of aconitase and elevation of citric acid levels. Less known is its role as a minor source of inorganic fluoride, about 3% of the labelled ^{14}C-F bond being released as $^{14}CO_2$ over a 4-day period. In rats such conversion was confined to the first hour after intraperitoneal administration and did not increase as the amount of fluoracetate injected was increased. The principal metabolites were fluoroacetoacetyl CoA and its products, fatty acids and cholesterol and the fluoroacetate derivative, fluorocitrate [GAL *et al.*, 1961]. Taking the best estimate of an acute lethal dose in humans as 5 mg/kg, the maximum dose which could be of concern from the point of view of fluoride release would be about 2 mg/kg. Thus, 140 mg of fluoracetate would yield about 1 mg of fluoride for an adult.

Further evidence for the splitting of fluoracetate comes from the demonstration that feeding 20 ppm of this molecule to rats for 12 weeks resulted in an increase in the femur fluoride level to twice that of controls. Since the controls were fed a diet with 5 ppm of fluoride it would appear that the fluoracetate provided an additional equivalent of 5 ppm of fluoride to these animals. Since fluoride represents 25% of the fluoracetate molecule it was present as 5 ppm and the degree of conversion would have been 100%. The possibility of the presence of contaminating fluoride ion in the fluoracetate should be rigorously excluded, however, before such a finding can be accepted. While 100% conversion of organic to inorganic fluoride is not to be expected in mammalian systems, bacteria are known to produce this conversion with greater efficiency than higher organisms. Equal amounts of fluoride ion and glycolate are produced during the reaction and the displacing OH is derived from water rather than atomspheric oxygen. Such conversion might occur, of course, in the intestinal tract leading to a lowered level of fluoracetate and increased inorganic fluoride, but the necessary conditions for such action have not yet been shown to be present. Virtually all sources of F cleaving bacteria are gram-negative ones derived from soil or river mud. The characteristics required are temperatures of 30 °C, or below, pH of 8–9 and frequently fluoracetate as the sole carbon source.

Trifluoroacetate. While monofluoroacetate is both toxic and minimally metabolizable as a source of fluoride ion, the trifluoracetate moiety is generally regarded as nontoxic and inert. The anesthetic halothane, under aerobic conditions, is metabolized to chloride ion and trifluoroacetate, neither of which produces toxic effects under ordinary conditions of use. Where anaerobic conditions prevail, additional products are formed which clearly are the result of removal of halogen atoms. When rats were anesthetized with halothane, with either 40 or 7% oxygen, a difference was noted in the plasma fluoride level. Under halothane with 40% oxygen or under 7% oxygen alone plasma levels were 1.1–1.8 μM/l. With anaerobic conditions and halothane, plasma fluoride elevations of 13–14 μM/l were observed. When the animals were allowed to recover by breathing room air again, a return to normal plasma levels of fluoride occurred after a few hours. The presence of oxygen therefore eliminated the reductive pathways and fluoride ion accumulation stopped [WIDGER et al., 1976].

Low oxygen tension also increases the toxicity of halothane by increasing metabolites capable of binding to microsomal lipids, an effect not directly related to fluoride release.

A very small amount of $^{14}CO_2$ derived from the trifluoromethyl group of halothane has been demonstrated, but this is not the direct result of trifluoroacetate metabolism. Rather the defluorination is due to mercapturic acid formation. Mercapturic acid formation is a type of reaction which requires the presence of reduced glutathione and one of several enzymes derived from liver homogenate supernatant, in this instance glutathione S-alkyl transferase. The reaction proceeds by the displacement of a halogen by the glutathione. Subsequent reactions cleave off the γ-glutamyl and glycine of the glutathione following which an N acetylation of the cysteine occurs [BOYLAND, 1971]. The presence of low oxygen tension is necessary to provide a source of reduced glutathione. The halogen-containing substrate is usually a hydroxylated metabolite, the elements of HF being removed. An hydroxylated intermediate resembling trifluoroethanol has been postulated as the halothane source of the fluoride ion. With sufficient oxygen tension this intermediate will ordinarily be oxidized to trifluoroacetate [VAN DYKE and WOOD, 1973].

Patients receiving halothane have been shown to have elevated fluoride plasma levels after 5 h of anesthesia. The normal level of 1.7 may at that time be elevated to 3.0 μM/l. The level returns to normal after 24 h.

Fluoroxene is another anesthetic ether containing a trifluoromethyl group. Metabolism of this agent produces mostly trifluoroethanol and its glucuronide along with some trifluoroacetate. Here too, mercapturic acid formation is responsible for the quite small defluorination, about 0.06%. Plasma fluoride levels of 3 μM/l have been reported which are returned to normal levels at 24 h [CREASSER and STOELTING, 1973].

The anesthetic agent with the greatest potential for releasing inorganic fluoride does not contain the trifluoromethyl group. Methoxyfluorane is a halogen-substituted ethylmethyl ether. The ether may be cleaved leading to an unstable dichlorodifluorethanol which is defluorinated to dichloroethanol. Rats with differing ability to liberate fluoride ion from methoxyflurane showed no differences in their liver enzyme concentrations [VAN DYKE and WOOD, 1973]. This suggests that such differences may be due to short-term considerations such as redox conditions rather than the potential to hydrolyze the fluorine atom. Increased fluoride storage in the long bones of rabbits with chronic exposure to methoxyflurane has been reported [SMITH, 1970].

Methoxyflurane anesthesia has been shown to produce very high elevations of human plasma fluoride. From normal levels of 1–2 μM/l values as high as 190 μM/l have been reported. These excessively high levels are accompanied by thirst and polyuria and are usually associated with extreme conditions, i.e. prolonged anesthesia with minimal premedication in obese patients. Levels of about 100 μM/l are also accompanied by signs of nephrotoxicity: creatinine elevation, hypernatremia and serum hyperosmolality.

At still lower elevations, which are the most common ones observed, i.e. 50–70 μM/l after 3–5 h, signs of renal impairment are not evident [DOBKIN and LEVY, 1972].

An estimate of the amount of fluoride ion released by these anesthetic agents can be made by multiplying the increase in plasma levels by the plasma volume.

Over a 5-hour period, the 3 anesthetic agents have been reported to elevate the plasma fluoride by 1 μM/l for two of the least metabolized agents, fluoroxene, and halothane, and by 50 μM/l for methoxyfluorane. Using a plasma volume of 3.15 liters and converting from μM to milligrams, the first two agents will yield a minimum of 0.060 mg and the last agent at least 3 mg of fluoride ion.

Aromatic Molecules

Aromatic compounds containing fluorine can be shown to release the halogen during mammalian metabolism but here, too, to a far smaller degree than in bacterial systems. Benzene and fluorobenzene have similar fates, both being recovered unchanged in the expired air to the extent of 40% of the dose. Hydroxylation of the fluorosubstituted form to a catechol occurs to the extent of 12%, the product 4-fluoroacatechol, reflecting resistance to the loss of fluoride. The chief source of defluorination is via glutathione mediated mercapturic acid formation, a reaction which accounts for only 1–2% of the total metabolism of the molecule.

With ring substitutions greater degrees of defluorination occur. o- and p-fluoronitrobenzene yield 40 and 33% mercapturic acid, respectively, the m-form yielding only 2% via this pathway. A similar difference occurs for the p-substitution with regard to oxidation to phenols, p-fluoronitrobenzene yields significant amounts of the defluorinated and reduced metabolite 4-aminophenol, while the o- and m-isomers give rise to only traces of this product.

From the above it follows that fluorine substitutions in the 4 position of aromatic rings are expected to be the most abundant source of inorganic fluoride due to metabolism. Phenylalanine with such a substitution gives rise to tyrosine by hydroxydefluorination, and an analogous reaction occurs for 4-fluoroaniline. Acetanilide with a 4-fluorosubstitution is less able to undergo the usual oxidation (94%) at that same carbon – but fluoride displacement by hydroxyls does occur to the extent of 22%. The remainder of the hydroxylations occur on the 2 (69%) and 3 (9%) positions. It is possible to predict, therefore, that if 4-fluoroacetanilide were to be used therapeutically under these conditions, its daily total of 800 mg of drug could provide as much as 20 mg of inorganic fluoride.

Fluoride atoms substituted in aromatic rings do not participate in the rearrangements described for other halogens under the rubric 'NIH shift'. Unlike other halogens fluorine atoms are removed rather than shifted to the adjacent carbon. As indicated above this is more common for *p*-substitution than for *o* and least common for m-substitutions. An interesting consequence of this homology is the relative toxicity of m-fluorotyrosine and the *o* and *p* forms of the fluorine-substituted amino acid. The former loses only a minimum of fluorine atoms and is converted to fluoroacetic acid, a metabolite with known toxicity. The *p* and *o* forms by virtue of defluorination reactions produce none of this metabolite and are significantly less toxic than the m-isomer.

Therapeutic agents which have incorporated fluorine atoms into aromatic rings include tranquilizers of the butyrophenone type. The prototype haloperidol contains a *p*-substituted fluorine in the butyrophenone phenyl ring. Metabolic products from haloperidol indicate that at least 86% of the drug is metabolized without fluorine removal. The 4-fluorophenyl moiety is found as a component of phenylacetic, propionic and benzoic acid metabolites and their conjugates with glycine [SMITH, 1970].

Antithyroid agents such as 3-fluorotyrosine have been shown to be toxic as measured by growth inhibition following ingestion for 1 month. NaF fed at a level which was 100 times the amount of fluorine in the tyrosine produced no growth inhibition. Likewise, the skeletal fluoride was not increased by feeding of fluorotyrosine even up to lethal levels. The above findings serve to confirm the fact that m-substitution produces negligible fluoride ion formation [SMITH, 1970].

Steroid Molecules

Fluorine substitution in steroids has produced a number of highly effective anti-inflammatory agents which are in common use today. Similarly fluorine substitution has found use in steroids having anabolic-androgenic effects. Both types of fluorine-substituted agents are in use clinically. The 9α-substitutions which are the chief modifications with fluorine produce no inorganic fluoride. The metabolic products identified for 9α-prototypes flurocortisol and fluorocortisone were 11β-hydroxy derivatives rather than the 11 ketometabolites usually found in the absence of the fluorine [BUSH and MAHESH, 1958]. The fluorine atom was intact in all 6 urinary metabolites found and identified.

Trifluoromethyl Groups

Trifluoromethyl groups have been added to a variety of agents as potentiators or modifers of specific properties. Therapeutic agents with CF_3 groups include phenothiazine tranquilizers and benzothiadiazine diuretics. Three examples of trifluoromethyl substitutions in the phenothiazine family are known, all with the same substitution. Intubating one member of this group trifluoperazine, to rats over 26 weeks at 1 mg/kg or 10 times that dose for 11 weeks failed to show any increase in femoral fluoride concentration [HORIKAWA et al., 1960].

Benzothiadiazines are diuretics of use in heart failure, renal disease or hypertension. Substitution of a CF_3 group on the aromatic portion of the molecule adds to the effectiveness of the agent.

The fate of the CF_3 was studied by feeding one of these compounds, bendroflumethiazide, by stomach tube for 85 days. Comparison was made with animals given NaF by intubation or in drinking water. Caries incidence was lowered for both the inorganic fluoride-fed animals and incisor depigmentation was visible in the intubated animals. The bendroflumethiazide-fed animals exhibited high caries and normally pigmented incisors. The daily dose of fluoride potentially available from the diuretic was 2 mg/day, the amount of inorganic fluoride which caused a reduction caries or incisor depigmentation was 0.9 mg/day [HASSELMAN and ROHOLT, 1963].

Fluoridated Water in Renal Dialysis

Chronic renal failure is associated with abnormalities of the skeleton which are a result of the impaired renal function. The skeletal lesions observed include excessive resorption which is roentgenographically visible and excess formation of osteoid which can be observed in microscopic sections. The severity of bone resorption can be shown to be correlated with high levels of immunologically reactive parathyroid hormone and the resorption can be seen roentgenographically when the dialysate Ca level is 5.7 mg/100 or less [CORDY et al., 1974]. Use of dialysates having higher Ca levels of 6-7.4 mg/100 is able to suppress the bone resorption of uremic osteodystrophy seen in these patients. This is achieved in this instance without producing hypercalcemia or increased calcinosis.

The second abnormal skeletal feature of renal dialysis subjects is the increased amount of osteoid formed on the surface of bone tissue. Here, the presence of 1 ppm fluoride in the dialysate increases the area of bone covered by osteoid [LOUGH et al., 1975]. The effect has been related to an abnormal stimulation of osteoblasts by the renal inadequacy which predisposes the cells to the action of fluoride. LOUGH *et al.* [1975], having studied this condition has reported that the osteoblasts are increased in number, the bone canaliculi are in disarray and the collagen fibers, while of normal periodicity, have an irregular alignment. Since fluoridated water has a fluoride level of about 50 μM/l it is obvious that the use of fluoridated water for the dialysis fluid is introducing an excessively large amount of the halogen for these patients. Removal of the fluoride prior to dialysis has been shown to provide a beneficial effect on the severity of osteomalacia observed [POSEN *et al.*, 1972]. The degree of osteoid coverage was reduced from 20 to 7.5% and the thickness of the osteoid seams was reduced from 38 to 17 μm in a group of 14 dialysis patients so treated. Since dialysis with fluoridated water takes longer than dialysis without fluoride the data were corrected for the duration of dialysis. Bone fluoride was reduced by $3 \times$ after the use of fluoride-free dialysate and a positive correlation between osteoid seam thickness and bone resorption was observed [JOWSEY *et al.*, 1972]. This latter fact is consistent with the observation that a double-blind study using NaCl and NaF in dialysis patients showed no difference in the osteoid seam index if Ca was added to the dialysate according to the protocol mentioned above [OREOPOULOS *et al.*, 1974]. The osteomalacia may also be relieved by administering 1.25 dihydroxylated vitamin D but not by vitamin D_2. This is in agreement with the requirement for adequate renal function to obtain the dihydroxylated form of the vitamin [EISMAN *et al.*, 1976]. It thus appears that removal of fluoride from dialysis fluid can reduce the risk of morphologic osteomalacia but the same results can follow if Ca levels of the dialysis fluid are raised above the level of 6 mg/100 ml. The role of fluoride in excess in producing osteomalacia during renal dialysis is thus not clearly established [POSEN *et al.*, 1972].

Is Fluoride Increasing in the Environment?

The desirability of fluoride as a preventive agent for dental caries does not require that unnecessary exposure to this element be tolerated. The most likely contribution of airborne fluoride has already been indicated to be below 0.04 mg/day. Any danger to human health from this source is confined to industrial workers exposed to much higher levels and more rarely to individuals in areas downwind from the emissions of these plants. Analyses of air samples do not support the contention that fluorides are increasing as part of a general pattern.

Food plants are generally not regarded as a major source of fluoride in the human dietary. Emissions from industrial plants have been shown to be of importance in grazing animals [SUTTIE, 1977], but the selectivity and preparation of vegetables used for humans reduces the exposure to extraneous fluoride to an insignificant level [HODGE and SMITH, 1977]. Differences in fluoride levels in vegetation exposed to fluoride containing emissions and those free of contamination may be as large as 10–20 times [OELSCHLAGER, 1970]. Evidence that plants so exposed are increasing annually in fluoride content is not compelling. JONES et al. [1971] found only marginal differences in leafy greens exposed to emissions in the midlands area of England from 1965 to 1969. Washing of vegetables reduced fluoride contamination by $1/3$ to $1/2$ of the unwashed values [JONES et al., 1971]. Blowing dust and cross-contamination have thwarted efforts to relate levels of fluoride in plants to distance from the site of emission. However, the areas of contamination are in general locally circumscribed and are ordinarily excluded from raising of food crops [JONES et al., 1971].

One possibility of increasing exposure to fluoride derives from the use of fluoridated water in the preparation of beverages from such water sources. The amount of beverage consumption may add fluoride to the diet of individuals beyond what is present in food or drinking water. This factor may be responsible for claims that fluoride in the dietary is increasing. Such assertions are almost always based on comparisons of present day ingestion with those reported by McCLURE [1943]. McCLURE did not adequately allow for any fluoride contribution in fluids except for drinking water in his calculations of the fluoride content of the human dietary. He used a tenfold range of possible fluoride levels and provided an estimate of 0.56 mg/day for the highest intake in the 10- to 12-year-old age group.

Subsequent balance studies or direct measurements of total dietary fluoride have generally shown higher intakes of fluoride on a daily basis.

A 3-day balance study of 23- to 24-year-old women by HAM and SMITH [1954] gave daily intakes of 0.43–0.76 mg. These diets, however, were devoid of fish or tea and included no additional soft drinks.

MARIER and ROSE [1966] analyzed food concentrations of fluoride. They included comparisons of food prepared with and without fluoridated water and sampled commercial beverages as well. In 7 male laboratory workers they found soft drinks and beer to be consumed at levels of 350 ml/day. Their own analyses of these beverages showed fluoride concentrations of 0.7 ppm in such beverages prepared in fluoridated water. Beer ranged from 0.1 to 0.3 when prepared in nonfluoridated water [McCLURE, 1949; OELSCHLAGER, 1970; MARIER and ROSE, 1966]. MARIER and ROSE's estimate of total fluid intake for the 7 subjects ranged from 1,020 to 3,150 ml/day.

The highest score represents an unusually high intake of total fluids. It is most probably not typical of an overall dietary pattern. The highest fluoride intake estimated in this study was for this same individual and was 5.0 mg/day. Others in the study had estimated total fluoride intakes of 4.0 mg/day or less.

SAN FILIPPO and BATTISTONE [1971] provided data on all beverages used in a fluoridated area (Baltimore). They found an average of 1.34 mg/day of fluoride contributed by all beverages. MARIER's data for total fluids gave a range of 1.1–1.8 mg with an average of 1.4 mg when the exceptional case mentioned above was eliminated.

LEE [1973] gave data for the fluoride content of all dietary items for 10 youngsters in Marin County, Calif. prior to fluoridation. The data include soft drinks made elsewhere with fluoridated water but do not include drinking water. By adding his analyzed total fluoride data to the amount of fluoride expected in drinking water at the respective age levels as given in table IIb, it should be possible to determine the total fluoride that would be ingested daily in the presence of fluoridation. These range from 1.6 to 2.0 mg/day. This is in fair agreement with the values of 1.0–1.7 mg given by MCCLURE [1943] and is inconclusive regarding the view that contamination by fluoride is increasing as part of a general pattern. The superficial appearance of greater fluoride intakes from food in this more recent survey is due in this instance to the inclusion of beverages made with fluoridated water in the dietary; an item not included in MCCLURE's calculations.

Emphasis so far has been on determinations of fluoride content of food and liquids. The inherent difficulty of these determinations has led to considerable study of alternate methods of assessing fluoride intake. The most reliable and often tested method is urinary analysis. It is possible to determine daily fluoride intake by analysis of 24-hour urine samples. In general the total ingested fluoride found in urine is about 50% under circumstances which the skeleton has not become saturated with fluoride. This includes those individuals who have been on steady intakes of fluoride for short periods, or those who have suddenly ingested a larger dose of fluoride than previously. Prolonged steady intake will yield urinary excretions of about 80–100%. This is shown in table VI which is adapted from LARGENT [1961]. It can be seen in this table that individuals consuming drinking water with F concentrations up to 8 ppm and having daily F total intakes of up to 15 mg, excrete from 75 to 100% of the fluoride ingested, with the most characteristic excretion of about 85% of the total. This so-called steady state is characteristic of long time residence in areas with constant water F concentrations.

Table VI. Balance studies of fluoride intake and urinary excretion (LARGENT, 1961)

Subject	H₂O, ppm	Dura-tion, years	Obs. period days	Fluids	mg F foods	total	urine	Percent excre-ted
1	2	10	96	2.42	1.17	3.59	2.86	80
2	2	14	96	2.55	0.94	3.49	2.88	83
3	5.5	29	60	3.81	1.32	5.13	4.49	88
4	5.5	48	61	8.14	1.35	9.49	8.32	88
5	6.1	14	160	6.34	1.98	8.32	7.76	93
6	6.1	34	133	6.74	1.02	7.76	8.09	104
7	8	19	140	12.4	3.13	15.5	12.9	83
8	8	19	140	11.3	2.47	13.8	10.4	75
9	20	8	45	20.8	1.48	22.3	12.3	55
10	20	11	65	15.6	1.16	16.8	11.4	68
11–19[1]	1	10	20	1.99	2.37	4.36	2.42	55

[1] SPENCER *et al.* [1969].

LEE has reported fluoride intakes of 0.911–1.518 mg exclusive of drink-ing water in unfluoridated Marin County in California. If these intakes were characteristic of the long-term fluoride intake of the community, one would expect that the urinary output represented 85% of the total intake. The reported urinary levels in a sample of 72 specimens in this community ranged from 0.14 to 0.50 mg/24 h. The mean was 0.24. If these figures represented 85% of the total intake, the range for daily fluoride *intake* would be 0.16–0.58 mg with a mean of 0.28. LEE's figures therefore do not indicate a steady state for fluoride intake. The urinary fluoride does, however, represent closer to 50% (15–33%) of the total dietary intakes measured by LEE, suggesting that these intakes have not been representative and are providing fluoride levels higher than the ones common to the community.

LEE has also argued that the recommended daily fluoride intakes pro-vided by the American Academy of Pediatrics Committee on Nutrition (1972) indicate that an increasing level of fluoride is occurring in the environment. This references gives a total figure of 0.5 mg for unfluoridated areas. The source of this reference gave maximum fluoride intakes as 1.5 mg/day with the average 0.5 mg/day [MACHLE, SCOTT and LARGENT, 1942]. The diet used in this study was stated to have avoided fish products, bones and salted food. LEE's data are based on a sampling in which fish products were well re-presented, and which elevated the usual fluoride daily intake as judged by urinary levels.

Aside from the representativeness of any diet which has been analyzed for fluoride, reference back to MACHLE [1942] for comparison of current and past fluoride intakes has tended to invite a highly selective approach to this pioneering publication. MACHLE et al. [1942] did give a range of 0.5–1.5 mg/day as the mean daily intake in their balance study. They also indicated that increased fluid consumption during the summer months increased the fluoride intake by a factor of 2. Food intake also varied from 0.83 mg to a maximum of 1.2, a factor of 0.5. The essence of MACHLE's article was not that only an intake of 0.5–1.5 mg of F/day is safe because a balance is established at that level, but that variations over an extended time period with highs and lows, did not alter the fact that a steady state had been established. It need not be emphasized that failure to note the high values that do occur in Machle's report can give a misleading conclusion regarding the levels of exposure to fluoride when comparing current levels to those of 35 years ago.

SPENCER et al. [1969] studied 10 adults from a fluoridated area in balance studies. During a 24-day control period these patients had an average fluoride intake from food of 2.37 mg/day. However, in the presence of a total fluoride intake of 4.36 mg/day, urinary excretion was 2.42 mg/day. Since these individuals had been consuming fluoridated Chicago water at the same level for several years the urinary elimination of fluoride should have been much closer to 85% of the intake than the approximately 50% reported. The elimination of 50% of a given dose of fluoride is characteristic of a sudden increase in fluoride intake not previously present [MACHLE, 1942]. The data of SPENCER et al. [1969] does not establish that the dietary fluoride of a fluoridated community is increasing but it suggests instead that an elevated level of exposure to fluoride was produced by the hospital diet. The excretion pattern of fluoride in SPENCER's study indicates that a total of 4.36 mg/day intake was higher than the subjects in the fluoridated area had been receiving. Other hospital diets in fluoridated areas were surveyed by KRAMER et al. [1974]. Dietary fluoride ranged from 1.73 to 3.44 mg/day and for water 0.6 to 1.27 mg/l. Examination of SPENCER's data reveals that the dietary fluoride was within the range of values surveyed, 1.99 mg. The fluoride consumption via water, however, was high, averaging 2.37 mg/day. This suggests that drinking water was consumed in amounts ranging between 2 and 3 liters/day. The only other liquid consumed as such in SPENCER's diets was 200 ml of fruit juice daily. The absence of soft drinks, beer, etc., from this diet probably was compensated for by the use of drinking water. It has already been pointed out that the various non-tap water sources of fluid have somewhat lower levels of fluoride than drinking water *per se* (table IIb). The elevated fluoride intakes in this study may reflect the fact that exclusive use of drinking water for fluids may increase fluoride intake over what it is in more varied diets. It is also useful to recall that in POT's [1974] study of office workers at a glass foundry, water intakes of 2–3 liters/day also produced renal excretions of 50% of the fluoride.

This is borne out in the study by the same authors in which average fluoride intakes of 13.72 mg/day as NaF were used followed by urinary excretions of 7.48 mg (54%). Similar results were found by OSIS *et al.* [1974b], intakes of 13–50 mg gave urinary excretions of 55%. Fluoride intakes from diet only in 1966 and 1969 were 1.5 and 1.8 mg, respectively, but urinary excretions were 2.9 mg/day for both years. This 'negative' balance was undoubtedly due to the absence of the fluoride component from water. Using the 50% rule, a total intake of 5.8 mg of fluoride would have been consumed. This would have required water consumptions of 4.0 liters or more daily. Using the 85% rule, water consumptions of 1.7–1.8 liters would have been consumed to provide daily total fluoride intakes of 3.2–3.6 mg/day for the assumed steady state. It is not possible to conclude that an increase in fluoride intake between the years 1966 and 1969 occurred from such findings.

It is also evident that hospital-prepared diets are somewhat higher in fluoride than random shopping basket selections in fluoridated cities. SAN FILIPPO and BATTISTONE [1971] found random diets to provide less than 1.0 mg of fluoride exclusive of liquids in fluoridated areas.

MARIER's study measured fluoride concentrations in food and estimated the food contribution to be 1–2 mg/day. There is some doubt as to the estimates provided by MARIER. Assuming a 3,000 kcal requirement for his subjects. The amount of food consumed can be estimated as 3,000/4.5 or 667 g/day. Since MARIER's data are expressed as ppm of fluoride, the presence of 1.0 ppm or less of fluoride in food would contribute less than 1 mg of fluoride/day. Since most of MARIER's fluoride analyses showed less than 1 ppm of this element in food the actual contribution of these foods would be closer to 1 mg than to the 2.0 mg he estimated.

We may conclude that convincing evidence for increasing concentrations of fluoride in food is lacking. Surveys cited to show this trend are either based on McCLURE's estimated data, have had high fish content, have used a hospital diet which for yet unexplained reasons is higher in fluoride than the market basket diet, have overestimated daily food consumption, or have utilized excessively large drinking water consumptions.

II. Dental Fluorosis

It should be obvious from the first chapter that efforts to generalize the long-term fluoride intake of individuals are fraught with difficulties. The day to day variability of diet, temperature, etc., make milligram dosages too uncertain to be of practical use in a program which attempts to utilize the beneficial effects of fluoride on developing teeth. From the foregoing, the major factor in fluoride intake is water-borne fluoride, which contributes also to dietary sources. As were our efforts in the past, our efforts in the predictable future will seek an optimum level which is compatible with the observed minimal toxic effects of fluoride in a community being served by that water, i.e., dental fluorosis.

Climate and Dental Fluorosis

Field studies of water consumption have demonstrated that its variations are dependent on mean maximum temperature. GALAGAN and VERMILLION [1957] demonstrated that temperature was an independent variable of the environment which influenced water consumption. From these findings GALAGAN and VERMILLION were able to establish an empirical relationship in the form of an equation which related ounces of water consumed per day, per pound of body weight, to the mean maximum temperature. The latter term was the average of 5 successive years in F°. The full empirical equation: oz H_2O/lb $= -0.038 + 0.0062$ F° has been found to apply over the temperature range 50–80 °F (10–27 °C). It has been shown to have less application to areas such as Edmonton, Canada, where the mean maximum temperature is below 10 °C. In such communities water ingestion appeared to be influenced by a large contribution made by the indoor environment [MCPHAIL and ZACHERL, 1965].

GALAGAN et al. [1957] after observing a high incidence of dental fluorosis in Arizona at 0.8 ppm of fluoride, studied the fluid intake in 455 children in two California communities. By compiling data taken at different times of the year, a graph was prepared which gave fluid intake (in oz/day/lb body weight) as a function of daily maximum temperature. From the graph the previous equation was derived, allowing calculation of water intake corresponding to a single average temperature (T_{max}). To determine 'optimal' fluoride intake, the water ingestion corresponding to the T_{max} of Maywood and Joliet, Illinois, were determined from the graph. These cities had 1.2 and 1.3 ppm of fluoride in their drinking water and had been previously shown to receive maximal benefit from fluoride as a systemic anticaries agent. The fluid intake of these cities as determined from their T_{max} was 0.34 oz. This converts to 1,530 ml/day for a 70-kg adult. To obtain the optimal ppm of a water supply, the relationship, ppm $F = \dfrac{0.34}{E}$ was used. E in this instance is the actual water consumption being considered. For cities with the same T_{max} as the above reference cities, the ppm of F would be 1.0 to allow a margin of safety. For communities with greater T_{max} volumes, E is correspondingly larger and the ppm term becomes less than 1.0. In communities with a lesser T_{max}, the relationship does not apply.

MINOGUCHI [1970] has carried this analysis further. Although GALAGAN found the fluoride ingestion from Maywood and Joliet water to be optimal, these intakes do not allow for those communities where fluoride in dietary constituents may be more abundant. In Japan, where a great deal of fish is eaten, the total fluoride intake may be greater than that presumed for Maywood and Joliet. To correct for this, MINOGUCHI derived a formula from graphic analysis which took into account the community fluorosis index (CFI). DEAN [1936] had previously determined that a CFI of 0.4 or less was of negligible public health concern. A new optimal water fluoride level could be defined as the maximal effective anticaries F concentration not causing a CFI greater than 0.4.

Using GALAGAN's and DEAN's data derived from two areas in the United States with average temperatures of 10 and 21°C, a plot of ppm of fluoride in the drinking water vs. Community Fluorosis Index may be made. The data permit two straight lines to be drawn by inspection, and the two lines, one representing the temperature range of 21 and the other 10°C, intersect at a point consisting of CFI 0.2 for the ordinate, and 0.4 ppm for the abscissa. By converting these converging lines into a right triangle further analysis becomes possible. Figure 1 which represents this relationship consists of: a hypotenuse, which is the line describing the change in CFI produced by increasing water fluoride at temperatures of 10°C; an altitude, which is the corresponding relationship at 21°C; and a base, which represents the range

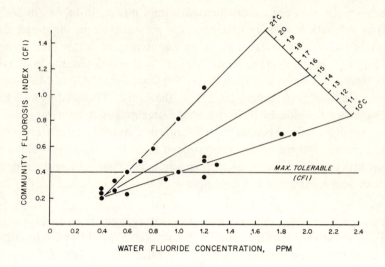

Fig. 1. Graphic method of obtaining optimal fluoride concentration.

of temperatures being compared, 11 °C. When read against the abscissa the base has units of ppm of fluoride; in this instance it has a span corresponding to 0.65 ppm of fluoride in the drinking water. By dividing the base into equal divisions a series of smaller right triangles can be obtained. The hypotenuses of these right triangles will cross the horizontal line drawn at CFI 0.4 at different points along the abscissa. These points of intersection can now be read as ppm of F, which values can be regarded as the maximal fluoride concentrations which will produce a CFI no larger than 0.4; this is the definition of optimal fluoride level given above. It is obvious that intersecting lines can be drawn for each temperature between 10 and 21 °C by dividing the base into eleven equal parts.

Table VII provides a comparison of three different ways of determining the optimal fluoride concentration in drinking water for temperatures in the range 10–21 °C. The column labelled Galagan is derived from the use of his trend formula, given previously. The column labelled Minoguchi is derived from his computations [MINOGUCHI, 1970]. The column labelled Graphic is derived from the method of analysis described above with the use of figure 1. The three methods predict comparable optimal fluoride concentrations in drinking water.

Table VII. Recommended fluoride concentrations in drinking water (ppm) in areas with average temperatures between 10 and 21°C

Average, °C	Galagan	Minoguchi	Graphic
10.0	1.0	1.1	0.97
12.75	0.82	0.87	0.82
15.5	0.71	0.75	0.70
18.75	0.64	0.67	0.64
21.0	0.58	0.61	0.57

Acceptable fluoride levels can also be determined, as they were in California, by using the appearance of moderate dental fluorosis as an unacceptable criterion [RICHARDS *et al.*, 1967]. No moderate fluorosis was seen with a water supply of 1.1–1.3 ppm in an area with a mean temperature of 18 °C. In areas of higher temperature moderate fluorosis was seen at 0.8–1.0 ppm, but not at 0.5–0.7 ppm. At still higher temperatures (27 °C or above) moderate fluorosis was observed at 0.5–0.7 but was not at 0.2–0.4 ppm. As will be seen in table VIII this method of using moderate fluorosis as a ceiling criterion is not a sensitive one because increasing intensity of fluorosis affects the milder forms of this condition before it has much effect on the moderate or severe forms. This is undoubtedly a function of the low frequency of the more severe forms of dental fluorosis in most populations. Table VIII provides a listing of studies of dental fluorosis conducted in various parts of the world. Severity of dental fluorosis is indicated by either the Community Fluorosis Index (CFI) or the percent of subjects presenting various degrees of fluorosis. The data derived from this broad base reveal that both fluoride concentration and temperature elevate the CFI. Another trend which emerges from table VIII is the more accelerated increase in the fluorosis categories 'very mild' and 'mild', as the fluoride concentration in drinking water increases. Although a minor increase in the 'moderate' category can be seen as ppm or temperature rises, most of the increased severity takes the form of a higher percent of individuals with the milder forms of fluorosis.

Areas of Excessive Fluorosis

Listed in table VIII are a number of studies which have shown an unexpectedly high level of fluorosis in the presence of water fluoride concentrations that were rather low, i.e. around 1.0 ppm. One such study in Lucknow,

Table VIII. Incidence and severity of dental fluorosis as influenced by climate and drinking water concentration of fluoride

Location	Average C°	H_2O F, ppm	CFI	Severity of fluorosis, percent of total					Percent afflicted	Reference
				doubtful	very mild	mild	moderate	severe		
Sweden	6	1.0	–	36	23	41	0	0	64	Forsman [1974a]
Sweden	6	5.0	–	8	0	65	5	20	90	Forsman [1974a]
Sweden	6	100	–	0	0	0	35	65	100	Forsman [1974a]
Austria	7	1.0	–	–	18	0	0	0	18	Binder [1973]
Austria	7	1.5	–	–	43	5	0	0	48	Binder [1973]
Austria	7	3.0	–	–	19	27	6	0	52	Binder [1973]
England	10.5	0.0	0.60	45	0	12	6	0	63	Forrest [1965]
England	10.5	0.0	–	–	4	1	2	0	47	Forrest [1965]
England	10.5	0.12	–	–	5	3	0.6	1.4	10	Goward [1976]
England	10.5	0.9	0.32	16	–	–	–	–	16	Forrest [1965]
England	10.5	1.0	–	9	3	0	0	0	12	Murray et al. [1956]
England	10.5	2.0	0.80	–	0	0	8	0	66	Forrest [1965]
England	10.5	3.5	1.9	–	–	–	–	11	92	Forrest [1965]
England	10.5	5.8	2.6	35	–	–	–	31	96	Forrest [1965]
United States	10	0.4	0.25	37	5	1	0	0	6	Galagan and Lamson [1953]
United States	10	0.5	0.22	35	4	1	0	0	5	Galagan and Lamson [1953]
United States	10	0.6	0.17	21	6	0.5	0	0	7	Galagan and Lamson [1953]
United States	10	0.7	–	3	2	0	0	0	2	Dean and Evonve [1937]
United States	10	0.9	0.31	35	10	2	0	0	12	Galagan [1953]
United States	10	0.9	–	21	9	2	0	0	11	Dean and Evonve [1937]
United States	10	1.2	0.32	32	14	1	0	0	15	Galagan and Lamson [1953]
United States	10	1.2	0.49	32	30	2	0	0	32	Galagan and Lamson [1953]
United States	10	1.2	0.51	28	29	4	0	0	33	Galagan and Lamson [1953]
United States	10	1.3	0.46	34	22	3	0	0	25	Galagan and Lamson [1953]
United States	10	1.5	–	20	18	6	1	0	25	Dean and Evonve [1937]
United States	10	1.6	–	8	22	4	0	0	26	Dean and Evonve [1937]
United States	10	1.7	–	21	37	5	0	0	42	Dean and Evonve [1937]
United States	10	1.8	0.67	32	30	9	0	0	39	Galagan and Lamson [1953]
United States	10	1.9	0.69	27	40	6	1	0	47	Galagan and Lamson [1953]
United States	10	2.5	–	14	28	22	14	3	67	Dean and Evonve [1937]
Italy	16	1.3	1.2	–	23	17	10	8	58	Massler and Schour [1952]

Country										Reference
Italy	16	3.5	2.5	-	5	38	53	4	100	Massler and Schour [1952]
Israel	20	2.0	0.80	-	-	-	-	-	57	Milgalter et al. [1974]
United States	20	0.4	0.21	32	2	1	0	0	3	Galagan [1957]
United States	20	0.5	0.30	38	9	1	0	0	10	Galagan [1957]
United States	21	0.6	-	-	+	+	0	0	-	Richards [1966]
United States	20	0.7	0.46	45	12	3	2	0	17	Galagan [1957]
United States	20	0.8	0.52	39	9	6	2	0	17	Galagan [1957]
United States	20	1.0	0.85	38	30	18	0	0	48	Galagan [1957]
United States	20	1.2	1.12	20	26	14	13	3	56	Galagan [1957]
United States	18	1.2	-	-	+	+	0	0	-	Richards [1966]
Morocco	20	0.4	2.2	6	20	29	29	13	91	Møller and Poulsen [1976]
Taiwan	24	0.0	0.10	-	11	4	0	0	15	Pu and Lilienthal [1961]
Taiwan	24	0.3	0.27	-	26	8	0	0	34	Pu and Lilienthal [1961]
Taiwan	24	0.7	0.44	-	40	14	5	0	54	Pu and Lilienthal [1961]
Taiwan	24	0.9	0.74	-	36	38	9	0	83	Pu and Lilienthal [1961]
Taiwan	24	1.6	1.02	-	21	37	24	2	84	Pu and Lilienthal [1961]
Pescadores	24	0.5	0.88	-	33	33	19	-	85	Pu and Lilienthal [1961]
Pescadores	24	0.5	0.51	-	35	22	5	1	63	Pu and Lilienthal [1961]
South Africa	25	2.4	-	-	15	27	41	-	83	Bischoff et al. [1976]
United States	27	0.3	-	-	+	+	0	0	-	Richards et al. [1967]
United States	27	0.6	-	-	+	+	+	0	-	Richards et al. [1967]
Australia	27	1.3	1.03	-	-	-	-	-	-	Kallis and Silva [1970]
Thailand	28	0.1	-	-	-	-	-	-	3	Leatherwood et al. [1965]
Thailand	28	0.1	-	-	-	-	-	-	42	Leatherwood et al. [1965]
Thailand	28	0.2	-	-	-	-	-	-	60	Leatherwood et al. [1965]
Thailand	28	0.6	-	-	-	-	-	-	56	Leatherwood et al. [1965]
Thailand	28	0.7	-	-	-	-	-	-	61	Leatherwood et al. [1965]
Uganda	29	0.2	0.04							Møller and Poulsen [1976]
Uganda	29	0.7	0.82							Møller and Poulsen [1976]
Uganda	29	2.5	1.74							Møller and Poulsen [1976]
India	32	0.2	-	-	11	1	0	0	12	Nanda et al. [1974]
India	32	0.6	-	-	18	5	1	0	24	Nanda et al. [1974]
India	32	1.0	-	-	18	16	7	0	34	Nanda et al. [1974]
India	32	>1.2	-	-	2.6	17	23	0	56	Nanda et al. [1974]

Table IX. Comparison of fluoride intake derived from drinking water in endemic and nonendemic dental fluorosis areas of Lucknow, India and North America (mg/day)

Age group	Endemic area	Nonendemic area	North America	Age North America
1	0.05	0.03	0.035	1
1–2	0.23	0.12	0.115	1–2
2–3	1.04	0.30	0.197	3–4
3–5	1.58	0.42	0.238	5–6
5–8	2.11	0.50	0.314	7–8

India [NANDA *et al.*, 1974] has compared an endemic fluorosis and a nonendemic area with regard to both water and fluoride intake. According to Galagan's formula, the safe level of fluoride for Lucknow's T_{max} would be 0.66 ppm. Table VIII indicates that this level was not satisfactorily safe. More intensive study at Lucknow has revealed some of the factors responsible for the high incidence of dental fluorosis. There are three seasons of the year in Lucknow which have varying temperatures as well as varying water fluoride concentrations. During the 6 months of hot dry summer, total intake of fluids may reach as high as 1 liter for children in the age group 5–8. The same is true of the 3 month monsoon season. During the 3 months of winter, fluid intake declines to 700 ml/day. These values are close to those found for Atlanta and other American cities where fluorosis is not common (table IIIa). In chapter I, it was previously pointed out that drinking water per se represented from $1/3$ to $1/2$ of the total fluid intake of children in the US and Canada. In India this figure is greater, ranging between 80 and 95% of the total. In the nonendemic area, F concentrations for summer, monsoon and winter seasons were 0.68, 0.65 and 0.41 ppm. These levels, as indicated before, would be either safe or borderline according to GALAGAN's rule. In the endemic fluorosis area the seasonal variation of fluoride concentration in water was 1.11, 1.10 and 0.98. Not only is a greater proportion of drinking water used in the both areas, but in the fluorosis area it contains fluoride levels which are higher than permissible in the temperature: fluorosis relationship.

Although it has already been emphasized that it is the fluoride concentration in drinking water which gives the most reliable correlation with the biological effects of caries reduction and dental fluorosis, it is instructive to compare the calculated fluoride intakes from drinking water per person in the endemic fluorosis and nonfluorosis areas of Lucknow.

Table X. Age of crown initiation, completion and eruption of human permanent teeth

Tooth	Calcification begins	Crown completed, year	Total years for completion of crown
Upper central incisor	3–4 months	4–5	3.67–4.75
Lower central incisor	3–4 months	4–5	3.67–4.75
Upper lateral incisor	1 year	4–5	3–4
Lower lateral incisor	3–4 months	4–5	3.67–4.75
Upper canine	4–5 months	6–7	4.6–5.67
Lower canine	4–5 months	6–7	4.6–5.67
Upper 1st premolar	1.5–1.75 years	5–6	3.25–4.5
Lower 1st premolar	1.75–2.0 years	5–6	3.0–4.25
Upper 2nd premolar	2.0–2.25 years	6–7	3.75–5.0
Lower 2nd premolar	2.25–2.5 years	6–7	3.75–4.5
Upper and lower 1st molar	at birth	2.5–3.0	2.5–3.0
Upper 2nd molar	2.5–3.0 years	7–8	4.0–5.5
Lower 2nd molar	2.5–3.0 years	7–8	4.0–5.5
Upper 3rd molar	7–9 years	12–16	3.0–8.0
Lower 3rd molar	8–10 years	12–16	2.0–6.0

The compilation of table IX provides the average F intake in mg for Lucknow children corrected for the seasonal fluctuation in both drinking water consumption and fluoride concentration. Included are comparative data from table IIb which provide the fluoride intake from drinking water only, in North America. It can be seen that for all ages the intake of fluoride from drinking water is less for North American children. At age 3, the fluorosis area children consume three times more fluoride than the non fluorosis area of India and five times more than North American children. At age 5–8 this differential is increased further.

The evidence provided here illustrates the rather small margin of safety which is characteristic of fluoride when comparing caries prevention to dental fluorosis. This margin of safety is critical during the period beginning at 2–3 years of age and extending past age 8. The certainty of existence of a critical age for the development of dental fluorosis is entirely a consequence of the development of the teeth. Kronfeld's Table of Calcification and Eruption of the teeth which has long been used to indicate the chronology of tooth crown formation is provided as table X [HILL, 1945]. The earliest mineralization of a permanent tooth crown occurs for the first molars, which complete crown formation at 2½–3 years. The teeth which most frequently display fluorosis are the canines, premolars and the upper incisors [POULSEN and MØLLER, 1974]. NANDA et al. [1974] reported the highest incidence of dental fluorosis in canines, premolars and second molars, with lower incidence being found in the incisors and first molars. Table X indicates that the age of crown completion for these most often involved fluorotic teeth are: canines 6–7 years, premolars 5–7 years, and second molars 7–8 years.

In addition, table X provides data on the total duration of crown formation. It can be determined from this chart that the canines and 2nd molars have the longest exposure to body fluids during formation of the crowns. (Third molars are usually not included in such considerations because of their high frequency of malformations.) Thus, the high incidence of fluorosis in the endemic areas of Lucknow can be attributed to the high intake of fluoride exactly during the period of formation of those teeth most susceptible to fluorosis.

Additional Sources of Fluoride

The importance of considering exposure to fluoride in great detail was well shown by ERICSSON and co-workers [1971, 1972, 1973] in a series of publications dealing with formula vs. breast feeding of infants. Starting with the observation that mothers milk was extremely low in fluoride (0.025 ppm) it could be shown by calculation that use of cow's milk diluted 1:1 with 1 ppm tap water could increase by 20 times the amount of fluoride received by infants given this regimen. An even greater exposure to fluoride occurs when dry milk formula is diluted 1:6 with 1 ppm tap water, the increase in this situation being about 50 times. The obvious question of whether such differences in fluoride exposure had any effect on the incidence of dental fluorosis was investigated. When breast-fed and formula-fed children were identified by use of a questionnaire given to parents, and the 8–9 year olds examined, very small differences in fluorosis in the teeth calcified during the first year of life were recorded. The fluorosis indices, as well as the number of 1st permanent molars and permanent upper incisors exhibiting fluorosis all showed a slightly higher score for formula-fed children, but the differences were not statistically significant. When the population of children was further subdivided into those who were fed exclusively by breast or by formula, the small differences persisted. The degree of fluorosis was mild, the opacities occupying about 25% of the exposed surface. The fluorosis indices were no different from those found previously in areas with the same water content of fluoride in Europe and the USA.

In a follow-up study, the amount of fluoride in the deciduous canines and molars of these children was measured, since much of the crowns of these teeth would have been mineralized during the first half year of life. Formula-fed children had only 2–3 times more fluoride in enamel and dentin than breast-fed children. This was true in spite of the much larger exposure to fluoride (50 ×), received by the formula-fed children. It was shown subsequently that the formula-fed infants did not excrete a greater proportion of F in feces than did breast-fed infants. Thus, discrimination against the high level of F in formula did not occur in the GI tract. Retention of fluoride in the formula-fed infants was about 50% of the total compared to 22% or less in the breast-fed infants. The additional fluoride obtained by the formula-fed infants must be assumed to have entered the growing skeleton with only a modest uptake by the mineralizing teeth. Cervical enamel and dentin had between 1 and 2 times higher fluoride levels than the occlusal portions, presumably because the occlusal portion of these teeth is mineralized prenatally. The relative resistance of teeth to fluorosis in spite of a 50-fold difference in intake of F in breast-fed and formula-fed infants must be attributed to a preferential uptake of F by bone. Calculations indicate that the maximum expected bone concentration due to the excess fluoride from formula feeding was 400 ppm. Osseous development in the bottle-fed children was actually more rapid than that of breast-fed children during the first year of life [MELLANDER et al., 1959]. The difference in fluoride supply was one of several factors which could have been responsible for this observation.

One often mentioned source of fluoride is the sediment in well water, which when consumed, can increase fluoride intake. In Lucknow this contribution was included in the fluoride intake data since it requires about 48 h for the fluoride to be removed by sedimentation. In Lucknow water was drawn daily and rarely stored more than 1 day.

Another source of fluoride is dietary, especially where fish is a major part of the diet. In some parts of the world such as the Pescadores islands of Taiwan, fish represents an unusually large portion of the diet; 1 lb/day per person being consumed including the bones [PU and LILIENTHAL, 1961]. Using the value of 15 ppm of fluoride for fish, the contribution of this food to the total fluoride intake can be as high as 1.37 or 1.5 mg per day per individual. Table VIII indicates that the incidence of fluorosis in the Pescadores islands is

higher than in comparable areas on Taiwan; and is in fact equivalent in intensity of fluorosis to areas in Taiwan with 1.6 ppm [Pu and LILIENTHAL, 1961].

Airborne dust was suspected by POULSEN and MØLLER [1974] as the
source of fluoride responsible for the high incidence of dental fluorosis in
Morocco, where phosphate mining is a major activity. This region was
using a water supply with 0.25–0.54 ppm F. Even allowing for the greater
consumption of water by the children in this area the incidence of dental
fluorosis was high. Earlier studies had sought to identify the malnutrition
common to this area as a causative factor. Caucasian children living in the
area had noticeably less fluorosis than the native children who received a far
more restricted diet [MURRAY and WILSON, 1948].

In a follow-up of their study in Morocco, MØLLER and POULSEN [1976]
were able to correlate the community fluorosis index with distance from the
airborne dust (2% fluoride) producing factory. As the distance from the
factory increased from 0.5 to 2.0 km, the CFI fell from 3.0 to 1.5 or less.
Further evidence for this origin of the fluorosis came from analysis of the
severity of fluorosis in children 10–14 years of age who were exposed to the
dust prior to the installation of filters by the factories. Children aged 7–9
who were exposed after the installation of filters had much less severe
fluorosis than the older group. These findings suggest the possibility that
caucasian children may have been living in areas remote from the factory
dust and that this factor rather than nutrition was responsible for the observed differences in the native and caucasian children.

Other studies of industrial dust emissions have related this source of
fluoride to the presence of dental fluorosis. HODGE and SMITH [1977] listed
five such surveys of children residing in areas 1–2 km downwind from the
source of emission. The airborne fluoride levels ranged from 0.03 to 14 mg
F/m^3 and could have provided from 0.6 to 280 mg/day. Comparisons of the
degree of fluorosis with control groups not exposed to the emissions revealed
increased severity of mottling ranging from 2 to 15 times.

Placental Transfer

The presence of fluoride in teeth and bones which are developed in utero
is an indication that no absolute barrier to fluoride exists in the placenta.
The high fluoride level reported for placental tissue is related to the areas of
calcification frequently found in this tissue [GEDALIA, 1970]. Both developing teeth and bones have a slight tendency to increase in fluoride content as

the age of the fetus rises from 6 to 9 months. In all such measurements, the teeth incorporate slightly less fluoride than either the femur or mandible in terms of ppm [GEDALIA, 1970]. The reason for this is not clear but may be related to rate of growth, vascularity and metabolic turnover rate [ARM-STRONG, 1970].

Fetal blood is as high in fluoride as maternal blood in 0.1 ppm fluoride areas. In 1.0 ppm areas maternal blood has a higher fluoride concentration than fetal blood and the same concentration as placental tissue. This has given rise to the idea that the placenta may act to filter out higher levels of fluoride by virtue of its high calcium phosphate content [GEDALIA, 1970]. However, the tendency of fetal skeletal tissues to increase in fluoride with age indicates that this action cannot be a controlling one. This is reinforced by the demonstration that the skeletal fluoride gradient with increasing fetal age applies to areas having water supplies of 0.1, 0.5 and 1.0 ppm [GEDALIA, 1970].

Nutritional Deficiency and Dental Fluorosis

One of the modifying factors in the occurrence of dental fluorosis is the possible predisposition for this condition caused by nutritional deficiencies in the sample population. The appearance of dental fluorosis in areas of the world where such deficiencies are thought to exist endemically has encouraged such conjecture. Among the deficiencies suspected are those most likely to be seen in an undernourished population: protein, vitamin C and calcium.

All these deficiency states have the common feature of skeletal hypoplasia and it is this aspect which first engages attention in assessing predisposition to dental fluorosis. Since the skeleton plays a major role in the distribution and fate of fluoride, faulty skeletal growth during the period of formation of the teeth would be expected to have a role in fluoride produced dental hypoplasia. The series of papers by ERICSSON and co-workers [1971–1973] have amply demonstrated the key role played by the growing skeleton in preventing the induction of dental fluorosis in suckling infants in spite of an exposure to fluoride of 20- to 50-fold greater in bottle and formula-fed regimens.

Nutritional deficiencies which reduce skeletal growth should affect the distribution of a given dose of fluoride which in turn should influence the presence or severity of dental fluorosis. Presumably, the effect is produced through elevation of fluoride during the extended period of skeletal growth.

Protein Deficiency in Fluorosis

Skeletal growth is reduced in the presence of protein deficiency and this nutritional inadequacy has been shown to produce a lesser uptake of fluoride by bone. Using monkeys [REDDY and NARASINGA RAO, 1971; REDDY and SRIKANTIA, 1971] fed NaF by stomach tube while at the same time providing them with a 2.5% casein diet. Control animals had a skeletal fluoride average level of 0.48% (dry fat-free bone) while protein deficient monkeys fed the same level of fluoride had 0.35%, a reduction of about 25%. An interesting aspect of this study was the fact that fluoride-supplemented animals (10 mg NaF/kg/day) retained as much or more Ca and P than did the control animals, an observation that has led to attempts to exploit high doses of fluoride in treating bone dyscrasias [SPENCER et al., 1969].

Avitaminosis C and Fluorosis

Another form of skeletal deficiency was observed in vitamin C deprived monkeys [REDDY and NARASINGA RAO, 1971; REDDY and SRIKANTIA, 1971]. In comparing daily intakes of 10–1.0 mg of vitamin C, REDDY found the deficient froup to have a lower fluoride level in the skeleton, in this instance 0.36%, a concentration similar in magnitude to the protein deficient animals. A similar improvement in Ca and P retention was also seen in the presence of fluoride in these animals. Any role of vitamin C deficiency in dental fluorosis would be expected to have a similar basis as that of protein deficiency by lowering the rate of bone formation, the amount of fluoride bound by the skeleton is reduced. While in this instance the vitamin deficiency was a factor in fluoride balance, other studies have failed to support these findings and make the role of avitaminosis C equivocal [STOOKEY, 1970].

Protein deficiency on the other hand, has been confirmed as a factor in the development of dental fluorosis [VAN RENSBURG, 1972]. After 1 month on a protein-deficient diet, a dose of 16 mg/kg of fluoride produced a much higher incidence of amelogenesis defects in rats than it did in the group fed an adequate protein diet. In this instance a single dose given subcutaneously was more toxic in the presence of presumed skeletal hypomineralization.

Calcium Deficiency and Fluorosis

Calcium deficiency can also be included in the list of skeletal hypoplastic factors. Here however the pattern is more complex. Aside from the skeletal deficiency produced by low Ca diets and the subsequent effect that this has on fluoride binding, Ca itself may react with fluoride and make it less available to the organism; a protective effect. The extent of such binding has been exaggerated, however, and is discussed in a subsequent section.

MASSLER and SCHOUR's [1952] observation that the severity of fluorosis in post World War II Italy was positively correlated with diminished body stature can be viewed in a new light from the above discussion. While they found that a scarcity of dairy products and therefore Ca was prevalent, the greater fluorosis could have been the result of protein deficiency.

ZIPKIN et al. [1959] utilized low phosphorus rickets to alter fluoride binding by the skeleton. The rachitic femora contained half as much fluoride as those of either the pair fed or *ad libitum* fed controls, 0.19 vs. 0.37 or 0.44%. Dental fluorosis was not directly evaluated by these authors but incisor fluoride concentrations were determined. No difference was found in this parameter, all three groups having concentration of 0.100–0.120%.

REDDY and NARASINGA RAO [1971] and REDDY and SRIKANTIA [1971] fed low Ca diets to monkeys in order to produce a skeletal deficiency in the presence of intubated NaF. Only a minor difference, if any, was found in the fluoride concentration of the bones. With femur control values of 0.48%, low Ca diets providing only 25% of control levels had fluoride concentrations of 0.46%. The fluoride concentrations were actually higher than controls if areas of the humerus containing newly formed osteophytic bone were included. A Ca high level to low level dietary ratio of 4 therefore produced a questionable diminution of skeletal fluoride.

A dietary Ca high to low ratio of 3, was used by HARKINS et al. [1963]. They were unable to show any difference in femur fluoride levels (0.30–0.34%) in the 3 dietary groups. They also found no change in incisor fluoride. That the animals receiving the higher Ca intakes had larger and better mineralized bones was demonstrated by the fact that the constant level of fluoride found in the bones of all groups, appeared to be lower in the better-fed animals if the data were expressed as milligrams of fluoride per gram of femur.

The above two studies indicate that Ca deficiency even at levels of 3–4 times less Ca than normal, cause no change in fluoride binding by the skeleton. The previous study by Zipkin demonstrates that even with a twofold alteration in bone fluoride caused by rickets, the presumed elevations due to unbound fluoride caused no difference in incisor fluoride uptake.

Ca Binding and Fluorosis

Assuming a fluoride level of 1 ppm, Feldman et al. [1957] calculated on the basis of activity and ion pair formation that Ca could bind from 0.03 to 2.8% of the fluoride depending on the hardness of the water. For Mg the figures were 10-fold higher. That Ca complexing of F in solution is quantitatively unimportant was shown by Largent and Heyroth [1949] with the demonstration that 96% of the fluoride in a solution of Ca F_2 was absorbed in an adult human. Since water hardness includes Ca levels of up to 100 ppm, it is extremely unlikely that water hardness alone could affect fluoride absorption from fluoridated water [Wagner, 1959]. The higher levels of Ca from dietary sources however requires separate consideration.

Weddle and Muhler [1954] added $CaCl_2$ to a solution of 2 mg of F given daily by stomach tube to rats. The calcium supplements ranged from 0 to 1.0% in tenfold increments. The presence of soluble Ca in levels as low as 0.01% caused a reduction in amount of fluoride found in femurs of from 0.54 to 0.44 mg. Further decreases were found at higher Ca concentrations: 0.22 mg F at 0.1% and 0.14 mg at 1.0%. Ca pyrophosphate at levels of 0.1 and 0.01% caused no reduction in uptake of fluoride by the femur. The authors concluded that levels of ionic calcium in solution can interfere with fluoride absorption and cite the fact that milk corresponds to a solution of 0.10% Ca. Mg and Al produced the same interference with fluoride uptake that was found for Ca. It should be realized that the Ca supplements in the above study were further modified in the GI tract by the Ca present in food – about 100 mg/day. Thus, the lowest dose of Ca was 110 mg (2.75 mM) the medium dose 200 (5 mM) and the highest 1,100 (27.5 mM). Since F was 2 mg (0.1 mM), the molar ratios of Ca:F ranged from 27.5 to 275.

Comparable to this was a balance study on 9 adult males in which Spencer et al. [1969] fed 1,448 mg of Ca daily and one of two different levels of fluoride over a 30-day period. The lower Ca:F molar ratio was 50 and the higher ratio was 160. For both regimens urinary excretion was 55–57% of the total dose of fluoride, indicating that absorption was not interfered with. Fecal fluoride was 11% at the high Ca:F molar ratio and 6% at the lower ratio. The data indicate that small differences in the availability of fluoride could be discerned when the Ca:F ratio in the diet was very high but the overall effect on fluoride absorption was negligible.

WEDDLE and MUHLER [1954] did not determine the urinary and fecal fluoride excretion of their rats. The carcass fluoride in those instances when no extra Ca was fed or when noninterfering Ca was administered (Ca pyrophosphate) was approximately 50% of the dose. With the ionic Ca supplements the carcass fluoride ranged only between 11 and 39%. The difference between these values and the 50% found where no interference was present could be in either the urinary or fecal fluoride. In the former situation the extra fluoride would have been absorbed, in the latter situation it would not have been.

That insoluble solid forms of fluoride are not well absorbed has been known for several decades. MACHLE et al. [1942] demonstrated that human volunteers receiving equal doses of 6 mg of fluoride absorbed 65% of a dose of NaF in drinking water, 41% powdered cryolite (Na_3AlF_6) and 29% of powdered bone meal.

GREENWOOD et al. [1946] maintained puppies on a diet of bonemeal and a commercial form of defluoridated rock phosphate for half a year. The animals received 5 mg of F/kg of body wt daily. Controls consisted of dogs fed NaF at the same level. To provide 5 mg/kg of fluoride using the bone and mineral supplements it was necessary to provide a diet with excessive Ca, 15–30 g/day. This represents about 5 times more than the recommended level for dogs, 3–7 g/day [ALTMAN et al., 1974; LONG, 1961]. The high to low dietary Ca therefore was about 5:1 when the powdered bone and mineral supplements were compared to the NaF-fed animals.

Analysis of the fluoride content of the bones after varying periods indicated that 25 days after start of the diet the NaF fed dogs has 2 times as much fluoride in their femurs. At 100 days the NaF animals had 10 times more fluoride. It is possible to interpret these findings by attributing the lesser uptake of fluoride to the high levels of calcium administered along with the fluoride of bone meal and rock phosphate. The high Ca can be presumed to interfere with the absorption of fluoride. Another interpretation seems more likely; the bone meal and rock phosphate powders are sufficiently insoluble to resist complete dissolution; passage through the gut occurred before the entire mass of mineral was dissolved. This latter interpretation is supported by a study previously cited [MACHLE et al., 1942] which indicated that all the fluoride of bone meal was not absorbed by animals fed the material, the efficiency of absorption being 2:1 in favor of NaF.

GREENWOOD's [1946] study did include observations on the state of the teeth of the dogs. Dental fluorosis was most severe with the NaF supplement and was equal to that of the nonfluoride control animals in the bone meal and rock phosphate supplemented dogs. Femur fluoride was 0.2% in the dogs exhibiting dental fluorosis and 0.020% in the unaffected animals. Non-fluoride control dogs had femur fluoride levels of 0.015%.

Fish protein concentrate (FPC) is another form of supplement which is high in fluoride. In humans the fluoride is only negligibly less available than the same dose of NaF. SPENCER et al. [1969] found 88% of the fluoride in FPC to be absorbed, compared to 94% from NaF.

Dietary fluoride is slightly less available than the same amount of fluoride consumed in water [WEDDLE and MUHLER, 1954].

In the human balance studies of SPENCER et al. [1969] referred to earlier two levels of fluoride intake were given, 13.72 and 4.38 mg/day. Dietary Ca was 1,400 and the Ca:P ratio was 1:1. At the lower fluoride intake the percent retained was 32% while at the higher fluoride intake 38% was retained. Thus, a variation of 3-fold in fluoride intake in the presence of a constant intake of Ca produced no inhibition of fluoride retention. In this same study the effect of fluoride on Ca absorption was assessed with the use of ^{47}Ca. No improvement of this parameter was found. In fact, slightly lower blood ^{47}Ca levels and slightly higher fecal ^{47}Ca levels were found with the higher intakes of fluoride.

The above survey of the role of high levels of Ca in retarding the uptake of fluoride and providing a protective effect against the occurrence of dental fluorosis indicates that the effect can occur. It was seen, however, only when the elevated Ca intakes were 5 times or more greater than in the control samples. From a clinical point of view such ratios are not likely to be seen in a field study. If such ratios occur at all it would be in the form of normal Ca intakes compared to deficient ones with $^1/_5$ as much Ca in the diet.

Given a recommended Ca requirement of about 600 mg/day [ABRAHAM et al., 1977], dietary Ca levels of 100 mg or less per day would need to be present in order to increase fluoride retention. Such conditions are not likely to arise in most populations. Possible exceptions are Uganda and Taiwan, and these are discussed below.

Since milk and its products are the chief source of Ca in most human dietaries it is of interest to compare the available Ca from this source in dietaries in the areas where fluorosis has been most prominent. In the USA Ca from dairy sources is about 800 mg/day/person. In the UK and Australia it is 700, Austria 672, Italy 580, Israel 400, India 132, Uganda 75 and Taiwan 13 [WALKER, 1972]. In Lucknow the total Ca intake per day for the critical age group of 3–8 years was 400–700 mg/child. When compared to the recommended intake of 600 mg/day, this represents a marginal deficiency at worst. In contrast, in Taiwan the very low Ca levels of the diet might have been influential in causing a greater uptake of fluoride by dental tissue and the development of dental fluorosis.

Physiological Assessment of Fluorosis

We can conclude from the previous section that dental fluorosis is often a product of a variety of exposures to soluble fluorides. The fluoride source can come from water, food, medicinals or industrial dusts and in all instances total fluoride intake is a combination of several of these. Given the above, it should be possible to assess the total exposure to fluoride by sampling blood, bone or urine. Until recently blood fluoride levels had been little used due to what was believed to be an inability to show increases in blood fluoride at the low levels characteristic of water fluoridation. Technical improvements such as the fluoride ion specific electrode, along with conceptual advances on the nature of blood fluoride have now made possible such an approach. This will be dealt with in a subsequent section.

Bone biopsy techniques are informative but of limited utility in dealing with a large human population. The obvious alternative of using teeth as representatives of the skeletal system has also been considered. However, as will be seen in a subsequent section, fluoride analysis of tooth enamel has provided data which are far too variable to utilize for diagnostic purposes.

Fluoride in Urine

Urinary analysis has long been recognized as the most reliable parameter for assessing exposure to fluoride. The steady state relationship referred to in table VI of chapter I has provided a useful approximation of fluoride intake. The tendency for a steady state or balance to be reached on a given intake of fluoride has been well documented. Table VI indicated that a steady state had been reached at the lower levels of fluoride intake after 10 years of exposure or less, while for the same time period steady states for the higher intakes had not yet been established. Table XI which is adapted from McClure and Kinser [1944], demonstrates a more direct relationship. For water fluoride levels between 0.5 and 5.0 ppm a distinct empirical relationship

Table XI. Relationship between urinary fluoride concentration (ppm) and drinking water fluoride concentration (ppm)

Water F ppm	Urine F ppm	Number of subjects	Reference
0.0	0.33	671	McClure and Kinser [1944]
0.1	0.30	85	McClure and Kinser [1944]
0.2	0.40	121	McClure and Kinser [1944]
0.5	0.60	26	McClure and Kinser [1944]
0.7	0.70	158	McClure and Kinser [1944]
1.0	0.90	573	McClure and Kinser [1944], Zipkin *et al.* [1956]
1.3	1.0	21	McClure and Kinser [1944]
1.7	1.5	87	McClure and Kinser [1944]
1.9	1.8	173	McClure and Kinser [1944]
2.0	2.0	16	McClure and Kinser [1944]
3.8	3.8	62	McClure and Kinser [1944]
4.7	4.3	68	McClure and Kinser [1944]
5.1	4.0	50	McClure and Kinser [1944]
5.8	4.0	129	McClure and Kinser [1944]

exists with that of urinary fluoride concentrations which allows either parameter in ppm units to predict the other. The most linearity and therefore reliability for such prediction lies in the range between 1 and 4 ppm in both water and urine. Brun *et al.* [1941] reported that fluorosed enamel was associated with 3 ppm of F in urine, and this would correspond to a water fluoride level of 3 ppm. By way of comparison cryolite workers in Brun's survey had urinary concentrations ranging from 2.4 to 43.4 ppm.

The above relationship did not require long-term residence in the community in order for numerical equality of water and urinary ppm to be established. Over a period of only days the urinary concentration reflected the drinking water concentrations in individuals who changed their drinking water sources [McClure and Kinser, 1944]. Observations made on populations in various parts of the world during the last 30 years have indicated that dental fluorosis can occur in individuals consuming water with less than 3 ppm of fluoride. Many of these observations can be explained as being due to elevated water consumption in areas of high average daily temperatures [Richards *et al.*, 1967; Nanda *et al.*, 1974]. Other reports have attributed the occurrence of dental fluorosis to dietary patterns involving high fish consumption [Pu and LILIENTHAL, 1961]. Studies in Italy [MASSLER and SCHOUR, 1952] and Thailand [LEATHERWOOD *et al.*, 1965] have attributed the occurrence to malnutrition as a predisposing factor.

McCLURE's relationship was for drinking water in moderate climates, not for total fluoride intakes. It must be concluded that the urinary/drinking water relationship described by McCLURE cannot be used with complete confidence in predicting the occurrence of dental fluorosis in areas where water and dietary patterns are significantly different from the US. It should also be realized that in establishing the relationship, the urine sampling involved the pooling of 20 ml samples of urine from individuals residing in areas with common fluoride exposure. Analysis was performed on 100 ml samples of these pooled urines which were collected between 9:00 a.m. and noon. It follows that fluctuations in a given individuals excretion of fluoride over a single day would not be easily detected by this method and that varying fluoride intakes during a 24-hour period would not become evident in this analysis. If dental fluorosis were related to transient short term elevations of blood (and urinary) fluoride, sampling would have to be made at short regular intervals over a 24-hour period, for a single individual, to detect these fluctuations. Such short term fluctuations in blood fluoride have been recorded by EKSTRAND *et al.* [1977] in individuals consuming 1.2 ppm water.

Fluoride Content of Skeletal Tissue

The affinity which exists between fluoride ions and the bone mineral prototype, hydroxyapatite, is well established. Not only is fluoride a bone seeker but it is cumulative throughout life. It should be possible to utilize bone levels of fluoride to determine the degree of exposure. This can be and has been done, but the biopsy technique would be hard to justify for large scale appraisal of dental fluorosis. Nevertheless, skeletal fluoride does represent a reliable guide to an individuals lifetime exposure to fluoride. A positive correlation of 0.53 exists between bone fluoride and age [PARKINS *et al.*, 1974]. A scale of skeletal fluoride concentration can be developed which will parallel the various levels of exposure to fluoride. Such a relationship is of use in questions concerning industrial toxicity in both man and animals. Table XII presents the percent of F in bone ash and dry weight at varying levels of exposure to waterborne fluoride. The data in table XII are obviously of some value in assessing exposure to fluoride in a quantitative

Table XII. Relationship of drinking water fluoride to bone fluoride

Water F ppm	Percent F in bone ash	Percent F dry fat free bone	Reference
0.1	0.08	0.04	ZIPKIN et al. [1958]
0.2	0.08	0.04	ZIPKIN et al. [1958]
0.3	0.13	0.07	ZIPKIN et al. [1958]
0.4	0.14	0.07	ZIPKIN et al. [1958]
1.0	0.30	0.15	ZIPKIN et al. [1960]
1.0	0.32	0.16	PARKINS et al. [1974]
2.0	0.35	0.17	BOISSEVAIN and DREA [1933]
2.6	0.50	0.25	ZIPKIN et al. [1960]
4.0	0.81	0.41	ZIPKIN et al. [1960]
8.0	1.11	0.56	McCLURE et al. [1958]

manner, but much of the scale involves exposures which are well above the threshold levels implicated for the dental condition.

ZIPKIN et al. [1958] has provided a simple linear relationship between F in bone and degree of exposure to fluoride. At drinking water fluoride concentrations between zero and 4 ppm, a simple relationship exists similar to the one described for urinary and water ppm by McCLURE and KINSER [1944]. Using iliac crest, rib, vertebra and sternum as bone samples, a linear relationship can be expressed for either percent F in dry fat free or percent F in ash of human bones.

In an early study of rat bones and teeth, fluorosis was not observed in the incisors if their fluoride content was less than 0.04% [McCLURE, 1939]. The bone fluoride content corresponding to this was 0.07% of the fat free dry weight. From table XII this would identify water fluoride levels of between 0.4 and 1.0 ppm as the maximum permissible to prevent dental fluorosis in humans. We conclude that the sensitivity of the rat incisor to fluoride is greater than that of human teeth, and that McCLURE's data should not be applied.

Taking the data directly from table XII, the threshold for mild dental fluorosis in humans is generally accepted as 2 ppm in the drinking water. This would suggest that a skeletal fluoride of 0.35% in ash wt or 0.17 in dry weight would be accompanied by dental fluorosis under appropriate conditions.

Predicting dental fluorosis from bone fluoride would only be possible if the exposure occurred during the formation of the teeth, a fact which cannot be established from bone fluoride levels alone. Furthermore, all indi-

Table XIII. Fluoride content of surface enamel obtained from communities with varying fluoride concentrations in drinking water (ppm)

Nonfluoridated	Fluoridated	Fluorosed	Reference
50–70	130		ELLIOTT and SMITH [1960]
100	130	350–650	McCLURE [1956]
600–1,250	900–1,550	2,000–3,300	BRUDEVOLD [1962]
850		1,550	KEENE [1974]
850–1,200			MØLLER *et al.* [1965]
1,440			KOCH and FRIBERGER [1971]
1,700	2,200–3,200		AASENDEN and PEEBLES [1974]
		1,750	BISCHOFF *et al.* [1976]
		5,000	WEATHERELL *et al.* [1977]
		3,00–14,500	LEX [1974]

viduals exposed to elevated levels of fluoride in drinking water at the appropriate time do not develop fluorosis. Table VIII indicates that it is unusual to find 100% of a population with dental fluorosis even when the drinking water fluoride levels are well above what could be considered a threshold.

An obvious alternative to skeletal data would be a compilation of fluoride levels in normal and fluorosed teeth or more specifically in dental enamel. Table XIII presents such data and reveals that while fluorosed enamel has higher fluoride concentrations than fluoridated or nonfluoridated specimens, the data have great variability. The overlap of values and ranges makes it impossible to identify any precise concentration range which will define dental fluorosis. The reasons for the high variability of enamel fluoride levels fall into four categories: having to do with preeruption fluoride uptake, distribution within the tissue, age and the role of attrition of the teeth.

Teeth are formed prior to eruption and therefore spend a significant amount of time in contact with body fluids of hematogenous origin prior to their emergence into the oral cavity. This is indicated in table X which provides the range of years required for completion of the crown and eruption into the oral cavity for each type of tooth. The differences by tooth type were used by AASENDEN *et al.* [1973] to relate the quantity of fluoride in the enamel to the length of time the tooth spends preeruptively in contact with interstitial fluid. Canines which have a longer preeruptive history than do incisors also have higher fluoride levels, in the same individual.

The largest single variable in fluoride levels of enamel is the pattern of distribution of the halogen in the tissue itself. A very steep gradient for fluoride exists from the outermost to the deepest layer of enamel. Early attempts to sample enamel for such analysis failed to show differences of sufficient magnitude between samples with known differences in exposure to fluoride. This was due to the diluting effect of the less concentrated fluoride in the deeper portions of the tissue. Subsequent sampling of the outermost portions of enamel have shown that the history of exposure to fluoride can be confirmed by analysis and that the surface concentration of fluoride is higher than that of the interior by a factor of ten times [BRUDEVOLD, 1962; WEATHERELL et al., 1977]. Sampling of enamel for fluoride analysis is now performed by biopsy techniques which allow a measure of depth of sample, usually this is based on weight of the sample removed. Needless to say, the fluoride content of a single specimen will vary with depth of the sampling. 40% of the variability of enamel fluoride is due to this uneven distribution [De PAOLA et al., 1974].

Age tends to increase the fluoride levels found in enamel [BRUDEVOLD, 1962] but this is further complicated by the phenomena of tooth abrasion and pathology.

The fourth factor which contributes to variability of fluoride levels in enamel is the demonstrable fact that the fluoride rich surface of enamel itself is not uniform in fluoride content. WEATHERELL et al. [1977] has demonstrated variations in distribution of fluoride from different portions of the surface layer. Some of this variation is due to abrasion or wear of enamel. Contrasting the fluoride content of the outermost surface layer of a non-abraded portion of enamel (cervical) with that of the incisal edge which is abraded LEX [1974], found a variation of 1.5–2.0 times more fluoride in the unabraded samples. Complicating this further, is the fact that high concentrations of F are found in enamel corresponding to areas of defective mineralization due to the high affinity shown by these areas for fluoride derived from ambient sources. These areas may be excluded in the sampling procedure if they are sufficiently visible, but visual inspection is not a reliable determinant [MYERS et al., 1952].

Plasma Fluoride Levels in Dental Fluorosis

Just as relationships have been found between urinary or skeletal fluoride and drinking water content of this element, it would not be unreasonable to expect plasma fluoride to reflect such exposure. That this has not been so is due to two factors which in turn are somewhat related to each other. The first of these is the low level at which fluoride exists in plasma, and the corresponding difficulty of performing such an analysis. Methods for fluoride analysis require removal of interfering substances and in many instances accuracy at the 1 μM/l level has been difficult to achieve without resorting to concentration of the fluoride. Diffusion has been the chief method of removing interfering substances. The general necessity of a 2-ml sample for such analyses requires the use of venipuncture and this in turn has inhibited efforts at multiple sampling. Very recently EKSTRAND [1977a] has developed a method for using small, 0.5 ml, samples of capillary blood. He has also adapted the 'known addition slope' technique using the fluoride electrode to assay fluoride levels down to 0.2 μM/l, with reasonably high accuracy. Other improvements in fluoride analysis have utilized extraction of F as fluorosilane into a non aqueous phase and subsequent reverse extraction into base to obtain fluoride ion in an aqueous phase virtually free from interfering substances [VENKATESWARLU, 1974].

Earlier studies of plasma fluoride were also inhibited by a concept, since shown to be erroneous. SINGER and ARMSTRONG [1960] measured plasma fluoride levels in individuals residing in areas with varied exposure to fluoride in drinking water. They were able to find no increase in plasma fluoride as drinking water fluoride increased until levels of 5 ppm were reached. They concluded that plasma fluoride failed to reflect increasing low level exposure because of a physiological homeostatic mechanism which quickly removed fluoride from plasma. According to this concept, only when the exposure to fluoride was high enough did the mechanism fail to keep plasma fluoride at constant levels. It was subsequently shown, that the analytical method used by SINGER and ARMSTRONG was actually measuring total fluoride and included in this total a form of fluorine which was not exchangeable (ionic). Ashing the plasma sample converted the non-ionic fluoride to ionic [TAVES, 1966]. The nonionic component is the major component when fluoride exposure is low and its level bears no relationship to fluoride ion intake in drinking water. Table XIV demonstrates that ionic or exchangeable fluoride contrary to earlier beliefs, does indeed increase as fluoride exposure increases. The presence of nonionic fluoride and its inclusion as plasma fluoride was in fact concealing the systematic variation of the smaller component. Table XV illustrates the apparent independence of nonionic fluoride from plasma ionic fluoride. GUY et al. [1976] have provided evidence which links the nonionic fluoride to a family of surfactant molecules which are used commercially to protect fabrics and leather and to make waxed paper and floor waxes.

Table XIV. Comparative plasma fluoride levels at different concentrations of drinking water fluoride

μM/l	Water source	Reference
0.21	nonfluoridated	HODGE and SMITH [1970]
0.32–2.2	0.05 ppm	COWELL [1975]
0.36	nonfluoridated	HENSCHLER et al. [1975]
0.38	0.1 ppm	GUY et al. [1976]
0.52	nonfluoridated	SMITH [1970]
0.70–1.7	0.1 ppm	BARNES [1968]
0.80	nonfluoridated	SINGER [1974]
0.84–2.9	1.0 ppm	HALL et al. [1972]
0.88	nonfluoridated	HAHNIJÄRVI [1974]
0.89	1.0 ppm	Guy et al. [1976]
1.0	nonfluoridated	HENSCHLER et al. [1975]
1.0	0.9 ppm	GUY et al. [1976]
1.0–2.6	1.0 ppm	SINGER [1974]
1.4–5.2	0.15 ppm	JARDILLIER and DESMET [1973]
1.6	1.0 ppm	POT and FLISSEBAALJE [1974]
1.8	1.0 ppm	DOBKIN and LEVY [1972]
1.9	2.1 ppm	GUY et al. [1976]
2.1	1.0 ppm	SMITH [1970]
2.1	1.25 ppm	EKSTRAND [1977a]
2.6	10 ppm (children)	FORSMAN [1974a]
3.0–14.6	3.8 ppm	JARDILLIER and DESMET [1973]
3.1	9.6 ppm (children)	EKSTRAND [1977a]
4.3	5.6 ppm	GUY et al. [1976]
5.2	10 ppm (adults)	FORSMAN [1974a]
5.2	9.6 ppm (adults)	EKSTRAND [1977a]

Table XIV indicates that plasma ionic fluoride plateau level may be as high as 2.0 μM/l in the presence of 1 ppm F in drinking water. It is not known what plasma level coincides with the appearance of dental fluorosis in humans but it is likely to be greater than 2.0. From this level, however, it is entirely plausible that transient peak fluoride levels as high as 4–5 μM/l, which are not detected in cross sectional surveys, could occur.

The significance of ionic fluoride in fluorosis is reinforced by the work of RUZICKA et al. [1974]. Using different forms of fluoride, they found that the degree of fluorosis correlated with the presence of fluoride ion rather than with complex forms of the halogen. With comparable levels of fluorine at 2.4 ppm in the diet and 150 μg of F/l in the drinking water, histological evaluation of rodent incisors indicated that monofluorophosphate, difluoro-

Table XV. Relationship of exchangeable fluoride to total fluorine in plasma

Exchange-able, μM/l	Non-exchangeable μM/l	Total μM/l	Exchange-able, %	Reference
0.38	1.2	1.58	24	Guy et al. [1976]
0.70	1.1	1.80	38	Venkateswarlu [1975]
0.80	1.0	1.80	44	Venkateswarlu [1975]
0.80	7.1	7.9	9.2	Singer [1974]
0.89	1.6	2.49	36	Guy et al. [1976]
1.0	1.3	2.3	43	Guy et al. [1976]
1.3	2.8	4.1	32	Venkateswarlu [1975]
1.3	2.8	4.1	32	Venkateswarlu [1975]
1.9	2.3	4.2	45	Guy et al. [1976]
2.1	5.3	7.4	28	Singer [1974]
4.3	1.1	5.4	80	Guy et al. [1976]

phosphate and hexafluoroaluminate produced much less dental fluorosis than NaF. In McClure's [1950] study fluosilicate was shown to produce the same degree of fluorosis as NaF when found in equal quantities in the skeleton. This complex salt, however, hydrolyzes in water to yield fluoride ion quantitatively. In Ruzicka's study monofluorophosphate which also is a source of fluoride ions gave the most comparable degree of fluorosis to NaF.

With the knowledge that ionic fluoride levels do reflect exposure to fluoride, the importance of the steady state fluoride level in plasma must be considered. Ekstrand [1977a] has shown that a steady state of plasma ionic fluoride can be detected at water concentrations of 0.25 ppm of fluoride. The consumption of water with 1.2 ppm produced, over a 36-hour period, a peak concentration occurring between the hours of 12 noon and 4:00 p.m. The peak values were about 2 μM/l while the low values were about 1 μM/l. With higher fluoride levels in the water (9.6 ppm) peak concentrations were again seen between noon to 4:00 p.m., but in this instance they were considerably higher. The height of the peaks appeared to depend on age. For 38-year-old individuals the peak values were between 4 and 6 μM/l, for age 17 they were 3 μM/l and for ages 10–14, 2 μM/l. The trough fluoride levels between peaks ranged from 1 to 2 μM/l.

Peak fluoride levels have also been demonstrated by administering doses of 1–10 mg of F as a single dose. Such peaks occur within 30 min of administration and decline with a half time of between 2 and 9 h [Henschler et al., 1975; Ekstrand, 1977a]. Here too, peak height depends on the size of the dose: 1 mg – 2.6 μM/l; 1.5 mg – 3.4 μM/l; 2.0 mg – 5.2 μM/l; 3.0 mg – 6.8 μM/l; 6.0–10.5 μM/l and 10 mg – 21.1 μM/l. A plot of these data with doses of F in mg as the abscissa gives a slope of 0.5.

With multiple dosing of 3 mg every 6 h, a series of peaks and troughs occurred with maxima of 6–7 μM/l and nadirs of 1–2 μM/l. Midpeak values were between 4 and 5 μM/l and the peak to trough ratio was 1:0.12, a factor of 8 [EKSTRAND, 1977a].

With a dose of 4.5 mg every 6 h, peak values in plasma were 11 μM/l and trough values were 1–2 μM/l. The mid-peak value was about 5.2 μM/l and the peak to trough ratio was 1:0.1.

From the above we can conclude that plasma fluoride fluctuations do occur as a result of dose schedule. With water fluoridation, dosing is both multiple and small, which tends to reduce the peak:trough height ratio. However, there is reason to believe that diurnal fluctuations are present even at low level fluoride intake [EKSTRAND, 1977a]. Plasma peaks were higher after 8:00 a.m. doses than after 8:00 p.m. doses and diurnal maxima occurred between 12 noon and 4:00 p.m.

Fluctuation of plasma fluoride in response to dose schedule takes on added significance in the light of studies which have compared constant and alternating doses of fluoride in producing dental fluorosis.

RUZICKA et al. [1973] administered the same absolute dose of fluoride to mice in drinking water, 0.9 mg/day. When this amount of fluoride was consumed in a concentration of 375 ppm, fluorosis was more severe than when it was consumed in a concentration of 150 ppm. When the 0.9 mg of fluoride/day was given as 150 ppm concentration it produced the same degree of dental fluorosis as only 0.36 mg per day given in the same concentration. The implications are clear, the concentration of fluoride rather than its absolute amount was the determinant of fluorosis induction. This would appear to contradict the relationship suggested earlier in this chapter, that greater volumes of relatively low fluoride water can account for the dental fluorosis observed in Arizona and Lucknow. RUZICKA et al. [1973] did not determine plasma fluoride levels, so it is not clear whether peak values of significant magnitude occurred in this parameter as a result of the change in fluoride concentration in drinking water, or as a result of change in volume.

ANGMAR-MANSSON et al. [1976] administered increasing concentrations of NaF in drinking water to rats for 1- to 2-week periods. At water fluoride concentrations of 20 and 30 ppm the incisor teeth showed definite signs of fluorosis which were not seen at 10 ppm. Oxytetracycline was used to mark the fractions of enamel formed during the different periods. By providing a feeding period of 45 min after overnight deprivation of food and water, the authors were able to obtain fasting and postingestion levels of plasma fluoride. Comparison of the postingestion and fasting levels indicated that peak elevations of fluoride in plasma following the feeding period were critical for the induction of dental fluorosis. Increased fluoride levels occurred at each increase of water fluoride concentration. At the 20 ppm level plasma fluoride concentrations reached 10 μM/l and the authors concluded that this value was the threshold for the development of dental fluorosis.

With blood samples obtained at weekly intervals, no conclusions could be made regarding the duration of the peak plasma values which accompanied the fluorosis, other than that the test period of 1 week at the 10 μM/l level produced fluorosis while a week at 6 μM/l did not.

SUTTIE et al. [1972] administered the same absolute dose of fluoride, 1.5 mg F/kg to young heifers over a 6-year period. All animals so dosed displayed some degree of fluorosis, but those animals who received the dose uniformly over the entire test period had lesser degrees of fluorosis than those who received the same total dose in an alternating pattern consisting of both higher and lower concentrations: 4 months at 3 mg/kg and 8 months at 0.75 mg/kg during each of the 6 years. The varying pattern of dosing produced higher fluoride levels in the incisors and higher plasma levels (50 μM/l) as well. The higher plasma levels persisted for at least the 4-month period of high dosing but fell during the 8-month period of low dosing. Plasma F did not fall below 15 μM/l during the lower dose period, however. This was undoubtedly due to the continued 0.75 mg/kg dose during this period, but it could also have been influenced by mobilization of skeletal fluoride.

Skeletal mobilization of fluoride did not occur in a parallel study of similar design however. In this investigation animals were fed 1.5 mg/kg for 6 months alternating with a 6-month period with a nonfluoride-supplemented regimen. The plasma fluoride levels during fluoride supplementation were 40 μM/l but they declined to normal control levels of 5 μM/l during the nonfluoride period. These findings support the concept that fluorosis is produced when plasma peak values exceed a critical level. For dairy cows, as for rats, this level appears to be 10 μM/l. It is not yet known for how long the threshold must be exceeded.

Additional evidence in support of a higher than threshold level being necessary for fluorosis can be gleaned from a study by HASSELMAN and ROHOLT [1963]. As control animals in a study of bendroflumethiazide metabolism they utilized two groups of rats. One was given 2 mg of NaF by stomach tube and the other was given the same dose in drinking water. At the end of the experimental period only the intubated animals showed dental fluorosis, a result compatible with the concept that a plasma fluoride peak value was produced by the all at once method of administration but not by the increments present in drinking water.

AASENDEN and PEEBLES [1974] compared a human population receiving fluoride as a prophylactic measure for caries. Those children who received the NaF as tablets had a twofold greater incidence of mild dental fluorosis than those receiving fluoridated water over the same period of time. A third group which received no more than 30 days total exposure to fluorides had the least incidence of mild dental fluorosis. Here too, the interpretation which best fits the findings is that an all at once dose of fluoride causes plasma fluoride to rise above a threshold level which is critical for the appearance of dental fluorosis.

Individuals who had been exposed to 10 ppm of fluoride in water and subsequently moved to low fluoride areas were found by FORSMAN [1974a] to exhibit 'delayed fluorosis'. In those teeth which were mineralized after the individuals had left the high fluoride area (2nd molars), mottling was found. It was less severe than in the teeth which mineralized earlier and the mottling exhibited a more uniform pattern than the other affected teeth. This was attributed to a mobilization of fluoride from the bones of these children after taking up residence in the low fluoride area. Whether a large enough peak fluoride plasma concentration could occur under these circumstances is open to some doubt. SINGER et al. [1976] has reported that only a small fraction of skeletal fluoride (8% after correction for new bone formation) was removed from the bones of rats in a 21-day period following a high fluoride diet. The percent fluoride in bone ash declined from 0.24 to 0.20 during this period with an accompanying initial rise in plasma fluoride from $9 \mu M/l$ to 13.5. The plasma elevation only persisted for 3 days.

SPENCER et al. [1975] reported that in humans, only 10% of the excess fluoride deposited in bones was released; all of which occurred in a 6- to 12-day period after discontinuing fluoride supplements. The authors attribute this finding, which differs from that of LIKINS et al. [1956], to the fact that the period of high fluoride intake was only 6 weeks in duration in contrast to the years of exposure present in LIKINS' study. This latter investigation followed the urinary fluoride excretion in Bartlett, Tex. after the water supply was changed from a concentration of 8 ppm to 1 ppm. Urinary fluoride was still at the level of 2–3 ppm rather than 1 ppm, 2 years after the water supply was defluoridated. The data for adults do show what could be interpreted as an abrupt fall of 50% of the initial level somewhere between 1 and 5 weeks after the new water supply was installed, but the data for children show only a gradual fall of urinary fluoride over the entire 2-year period. While no large pulse of fluoride enters the plasma during fluoride mobilization from bone, some elevation does appear to occur. The magnitude of the elevation and its duration have not yet been shown to be sufficient as well as present at the appropriate time to be responsible for the presence of dental fluorosis.

Summary Statement

Any inquiry into fluorosis is soon confronted with the 50-year of history of fluoride as a public health concern. Initially a suspect in the etiology of what was at the time an unknown disturbance of mineralization, fluoride has been in the center of the dental research arena for half a century. Dental fluorosis can be regarded as perhaps the best example of a completely preventable disease of the teeth. At the present writing it also shares with tetracyclines the distinction of being a known specific cause of dental hypoplasia. Proper control of these two agents are, sad to say, the only adequately documented forms of prevention which can be practiced in the interest of dental health. Proper control means not only restriction but proper use. It is important, therefore, to assure ourselves that fluoride as a preventive against caries is being used under optimal conditions which will not be counter productive. The tendency to search for innovations which would inadvertently increase exposure to fluoride is one example of such counter productive activity.

The presence of a constant level of plasma fluoride at low (0.25 ppm) fluoride intakes has been reported by EKSTRAND [1977a]. Such a steady state may be analytical in origin since water borne fluoride at 1.2 ppm will produce peaks and troughs if individuals are followed over a large part of 1 day. A single high dose of fluoride will cause an elevation of blood fluoride to a level of 20 μM/l. An increase of such magnitude is clearly not germane to water-borne fluoride. Since water is ingested in small increments, the dose of fluoride provided by each increment may not be detectable in plasma. At issue are occasional doses such as 0.3 mg of F derived from vitamin pills or dentifrices which though small, represent as large a single dose of fluoride as that acquired from 1 glass of drinking water.

A well-established concept of pharmacokinetics is the 'plateau effect', a term given to the apparent establishment of a constant plasma concentration of a drug which has been administered over a period of time. The so called

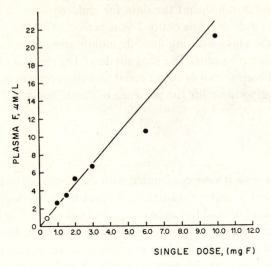

Fig. 2. Relationship of plasma fluoride maximum to single dose of F (mg). ● = Analyzed [EKSTRAND, 1977a; HENSCHLER *et al.*, 1975]. O = Extrapolated; peak concentration = 1.0 μM/l, plateau concentration = 1.5 μM/l.

plateau is actually made up of a series of peaks and valleys and the 'plateau level' is merely the concentration midway between these fluctuations. Table XIV provides a cross-sectional survey of the plasma fluoride levels found under varying water concentrations. The range 0.9–2.1 includes all the measurements dealing with fluoridation. Because of its random nature it is reasonable to regard this range of approximately 1.0–2.0 μM/l as the fluctuation due to the peaks and valleys and the value of 1.5 as the 'plateau' or midpoint level.

Knowledge of the plasma plateau level of fluoride characteristic of drinking fluoridated water can be used to determine the milligrams of fluoride per dose required to obtain such a plateau. 'A rule of thumb' generalization based on kinetic analysis has been developed. The plateau concentration after four or more doses of a drug equals 1.5 times (actually 1.44) the maximum concentration obtained after the same single dose of that drug. This relationship is based on the assumption that the dose frequency used in establishing the plateau is equal to the plasma half-life of the drug. Using this relationship, the plasma concentration of that single dose of fluoride capable of producing a plateau concentration of 1.5 is 1.0. Figure 2 provides a means for converting this 1.0 μM/l into milligrams. Figure 2 is a plot of plasma concentration vs. single dose of fluoride in mg derived from data provided by EKSTRAND [1977a] and HENSCHLER *et al.* [1975]. The linear relationship seen at low mg doses of F in figure 2 can be used to extrapolate the dose in milligrams corresponding to a plasma peak value of 1.0 μM/l. This is 0.35 mg. The above indicates that water fluoridation provides the equivalent of a dose of fluoride of 0.35 mg at a dose frequency equal to the plasma half-life. This latter was given by EKSTRAND [1977a] as between 2 and 9 h.

It becomes somewhat more obvious that supplemental doses of fluoride via food, medication or dust can have an effect on plasma fluoride even though they are present in sub-milligram amounts. A dose of 0.3 mg is sufficient to double the individual dose of fluoride provided by fluoridated drinking water.

One further consideration of the plateau concentration is indicated. Fluctuation of highs and lows around the midpoint concentration represented by the plateau bears a relationship to dose frequency. The greater the interval between constant doses the greater the size of the fluctuation from high to low. When dose frequency equals half-life, the ratio of concentration of high to low is 2. The ratio found in table XIV, which is also 2, is undoubtedly a fortuitous occurrence and does not imply that dose frequency is equal to plasma half-life in this instance. Nevertheless, changes in dose frequency, such as more frequent drinking episodes will have the effect of increasing the ratio between high plasma concentrations to low ones. The same effect will be produced by increases in the volume of water consumed since this will change the estimated equivalent dose of 0.35 mg to a larger figure. The effect of either increased frequency or increased volume of drinking fluoridated water will thus have the effect of magnifying the size of the fluctuation of the plasma levels. While none of these small increases in consumption can raise plasma levels to 10 μM/l their combined effect could be one of periodic transient increases to levels close to or above threshold concentrations. This would explain the random nature of the occurrence of dental fluorosis. Only those individuals will acquire the condition whose fluoride intake is sufficiently supplemented to cause such periodic elevations of plasma maxima over a period of time. Supplementation is far more likely in areas of high ambient temperature and therefore more individuals in these areas exhibit dental fluorosis. When viewed as above the larger absolute dose of fluoride consumed by those in warmer regions translates into higher frequency and higher individual doses, both of which have the effect of intensifying the magnitude of the fluctuation in plasma levels of fluoride.

The surprising degree of resistance to dental fluorosis shown by very young infants to high fluoride intakes [ERICSSON and RIBELIUS, 1971; ERICSSON et al., 1972; ERICSSON, 1973] has been explained as being due to the avidity of the growing skeleton for fluoride. In order for such protection of the teeth to take place the skeletal affinity for fluoride must be able to reduce either the height of the fluoride plasma peak observed or its duration. EKSTRAND [1977b] has reported lower plasma fluoride peak values in younger individuals residing in a 9.6 ppm fluoride area, an observation which is consistent with the above. This was also observed by FORSMAN [1974a].

We may conclude that our present knowledge of the relationship between fluoride intake and dental fluorosis indicates that: (1) For a given small absolute dose, of a few milligrams, the rate of administration is more important than the amount. (2) Fluctuations (highs and lows) around a plateau occur for fluoride as they do for other agents. (3) The occurrence of fluorosis due to high volume consumption of relatively low fluoride concentration water can be ascribed to changes in frequency of administration and individual dose and their effect on kinetics rather than to total dose. (4) The margin of safety for dental fluorosis compared to optimal fluoridation is small and is probably between 2 and 5. It is undoubtedly larger than this in very young infants because of their active skeletal metabolism. (5) Nutritional deficiency may predispose to dental fluorosis by an effect on the skeleton. Protein deficiency reduces bone growth and as a consequence permits higher than usual fluoride levels to prevail. Ca deficiency also reduces bone growth, but this is not accompanied by diminished fluoride uptake in the skeleton. Large amounts of Ca can interfere with fluoride absorption but only when the excess of Ca is 5 times or more greater than control levels. In order for Ca deficiency to promote dental fluorosis the degree of deficiency would have to be severe; in the order of $^1/_5$ of the recommended intake. (6) Every effort should be made to reduce unnecessary daily exposure to fluoride. Small supplements of fluoride of small benefit, which may cause transient elevations of plasma concentrations should be discouraged.

References

AASENDEN, R.; MORENO, E.C., and BRUDEVOLD, F.: Fluoride levels in the surface enamel of different types of human teeth. Archs oral Biol. *18:* 1403–1410 (1973).

AASENDEN, R. and PEEBLES, T.C.: Effects of fluoride supplementation from birth on human deciduous and permanent teeth. Archs oral Biol. *19:* 321–326 (1974).

ABRAHAM, S.; CARROLL, M.D.; DRESSER, R.D., and JOHNSON, C.L.: Advance data *6:* 3 (1977).

ALTMAN, P.L. and DITTMER, D.S.: Biology data book, vol. III; 2nd ed., p. 1454 (Biol. Hndbk. FASEB 1974).

Am. Acad. Pediatrics Committee on Nutrition: Fluoride as a nutrient. Pediatrics, *49:* 456–459 (1972).

ANGMAR-MÅNSSON, B.; ERICSSON, Y., and EKBERG, O.: Plasma fluoride and enamel fluorosis. Calcif. Tiss. Res. *22:* 77–84 (1976).

ARMSTRONG, W.D.: Fluoride and human health, pp. 134–136 (WHO, Geneva 1970).

AUERMANN, E. and BORRIS, W.: Fluoride intake in individuals in various age groups and under special conditions. Caries Res. *5:* 11 (1971).

BARNHART, W.E.; HILLER, L.K.; LEONARD, G.J., and MICHAELS, S.E.: Dentifrice usage and ingestion among four age groups. J. dent. Res. *53:* 1317–1322 (1974).

BINDER, K.: Comparison of the effects of fluoride drinking water on caries frequency and mottled enamel in three similar regions of Austria over a ten year period. Caries Res. *7:* 179–183 (1973).

BIRKELAND, J.M. and LØKKEN, P.: The pharmacokinetics of fluoride in a mouth rinse as indicated by a reference substance (^{51}Cr EDTA). Caries Res. *6:* 325–333 (1972).

BISCHOFF, J.I.; VAN DER MERWE, E.H.M.; RETIEF, D.H.; BARBAKOW, F.H., and CLEATON-JONES, P.E.: Relationship between fluoride concentration in enamel, DMFT index and degree of fluorosis in a community residing in an area with a high level of fluoride. J. dent. Res. *55:* 37–42 (1976).

BOISSEVAIN, C.H. and DREA, W.F.: Spectroscopic determination of fluorine in bones and teeth and other organs in relation to fluorine in drinking water. J. dent. Res. *13:* 495–500 (1933).

BOYLAND, E.: Mercapturic acid conjugation; in BRODIE and GILLETTE Handbook of experimental pharmacology, Chap. 55, pp. 584–608 (Springer, Berlin 1971).

BRUDEVOLD, F.: Chemical composition of the teeth in relation to caries. II; in SOGNNAES, Chemistry and prevention of dental caries, p. 50–51 (Thomas, Springfield 1962).

BRUN, G.C.; BUCHWALD, H. und ROHOLM, R.: Die Fluorausscheidung im Harn bei chronischer Fluorvergiftung von Kryolitarbeitern. Acta med. scand. *106:* 261–263 (1941).

BUSH, I.E. and MAHESH, V.B.: Metabolism of 9 α-fluorocortisone and 9 α-fluorocortisol. Biochem. J. *69:* 9 (1958).

COOK, H.A. and FRANCE, B.A.: Fluoride and tea. Lancet *i:* 1419 (1976).

CORDY, P.E.; GAGNON, R.; TAVES, D.R., and KAYE, M.: Bone disease in hemodialysis patients with particular reference to the effect of fluoride. Can. med. Ass. J. *110:* 1349–1353 (1974).

COWELL, D.C.: The determination of fluoride ion concentration in biological fluids and in the serum and urine of fluoride treated patients with pagets disease and osteoporosis. Med. Lab. Technol. *32:* 73–89 (1975).

CREASSER, C. and STOELTING, R.K.: Some inorganic fluoride concentrations during and after halothane, fluroxene and methoxyfluorane anesthesia in man. Anesthesiology *39:* 537–540 (1973).

DEAN, H.T.: Chronic endemic dental fluorosis (mottled enamel) J. Am. med. Ass. *107:* 1269–1272 (1936).

DEAN, H.T. and EVONVE, E.: Further studies on the minimal threshold of chronic endemic dental fluorosis. Publ. Hlth. Rep., Wash. *52:* 1249–1264 (1937).

DePAOLA, P.F.; BRUDEVOLD, R.; AASENDEN, R.; MORENO, E.C.; FOLEY, S.; BAKHUS, Y.; BOOKSTEIN, F; WARRAM, J.; ENGLANDER, H., and MEYERS, R.: The relationship of dental caries and enamel fluoride. J. dent. Res. *53:* abstr. No. 433 (1974).

DESHPANDE, S.S. and BESTER, J.F.: Absorption and retention of fluoride from ingested stannous fluoride dentifrice. J. pharm. Sci. *53:* 803–807 (1964).

DOBKIN, A.B. and LEVY, A.A.: Blood serum levels with methoxyflurane anesthesia. Can. anesth. Soc. J. *20:* 81–93 (1972).

EISMAN, J.A.; HAMSTRA, A.J.; KREAM, B.E., and DeLUCA, H.F.: 1, 25-Dihydroxy vitamin D in biological fluids. A simplified and sensitive assay. Science *193:* 1021–1023 (1976).

EKSTRAND, J.: Relationship of fluoride in the drinking water and the plasma fluoride concentration in man. Caries Res. (in press, 1977a).

EKSTRAND, J.: A micromethod for the determination of fluoride in blood plasma and saliva. Calcif. Tiss. Res. (in press, 1977b).

EKSTRAND, J.; ALVAN, G.; BOREUS, L.O., and NORLIN, A.: Pharmacokinetics of fluoride in man after single and multiple oral doses. Eur. J. Pharmacol. (in press, 1977).

ELLIOTT, C.G. and SMITH, M.D.: Dietary fluoride related to fluoride content of the teeth. J. dent. Res. *39:* 93–98 (1960).

EMMERSON, J.L. and ANDERSON, R.C.: Metabolism of trifluralin in the rat and dog. Toxicol. appl. Pharmac. *9:* 84–97 (1966).

ERICSSON, Y.: Fluorides in dentifrices. Investigations using radioactive fluorine. Acta odont. scand. *19:* 41–77 (1961).

ERICSSON, Y.: Effect of infant diets with widely different fluoride content on the fluoride concentrations of deciduous teeth. Caries Res. *7:* 56–62 (1973).

ERICSSON, Y. and FORSMAN, B.: Fluoride retained from mouth rinses and dentifrices in preschool children. Caries Res. *3:* 290–299 (1969).

ERICSSON, Y.; HELLSTROM, I., and HOFVANDER, Y.: Pilot studies on the fluoride metabolism in infants on different feedings. Acta paediat scand. *61:* 459–464 (1972).

ERICSSON, Y. and RIBELIUS, U.: Wide variations of fluoride supply to infants and their effect. Caries Res. *5:* 78–88 (1971).

FANNING, E.A.; CELLIER, K.M.; LEADBEATER, M.M., and SOMERVILLE, C.M.: South Australian kindergarten children. Fluoride tablet supplements and dental caries. Aust. dent. J. *20:* 7–9 (1975).

FELDMAN, I.; MORKEN, D., and HODGE, H.C.: The state of fluoride in drinking water. J. dent. Res. *36:* 192–202 (1975).

FORREST, J.R.: Mottled enamel. Br. dent. J. *119:* 316–319 (1965).

FORSMAN, B.: Dental fluorosis and caries in high fluoride districts in Sweden. Community dent. oral. Epidemiol. *2:* 132–148 (1974a).

FORSMAN, B.: The caries preventing effect of mouthrinsing with 0.025% sodium fluoride solution in Swedish children. Community dent. oral Epidemiol. *2:* 58–65 (1974b).

FULL, C.A. and PARKINS, F.M.: Effect of cooking vessel composition on fluoride concentration. J. dent. Res. *54:* 192 (1975).

GAL, E.M.; DREWES, P.A., and TAYLOR, N.F.: Metabolism of fluoracetic acid 2 C^{14} in the intact rat. Archs Biochem. *93:* 1–14 (1961).

GALAGAN, D.J. and LAMSON, G.G.: Climate and endemic dental fluorosis. Publ. Hlth Rep., Wash. *68:* 497–508 (1953).

GALAGAN, D.J. and VERMILLION, J.R.: Determining the optimum fluoride concentrations. Pbl. Hlth. Rep., Wash. *72:* 491–493 (1957).

GALAGAN, D.J.; VERMILLION, J.R.; NEVITT, G.A.; STADT, Z.M., and DART, R.E.: Climate and fluid intake. Publ. Hlth Rep., Wash. *72:* 484–490 (1957).

GEDALIA, I.: Distribution in placenta and foetus; in Fluorides and human health, pp. 128–134 (WHO, Geneva 1970).

GOLAB, T.; HERBERG, R.J.; PARKER, S.J., and TEPE, J.B.: Metabolism of trifluralin in carrots. J. agric. Fd Chem. *15:* 638–641 (1967).

GOWARD, P.E.: Enamel mottling in a non fluoride community in England. Community dent. oral. Epedemiol. *4:* 111–114 (1976).

GREENWOOD, D.A.; BLANEY, J.R.; SKINSNES, O.K., and HODGES, P.C.: Comparative studies of the feeding of fluorides as they occur in purified bone meal powder, defluorinated phosphate and sodium fluoride in dogs. J. dent. Res. *25:* 311–326 (1946).

GUY, W.S.; TAVES, D.R., and BREY, W.S., jr.: Organic fluorocompounds in human plasma. in FILLER (ed.) Prevalence and characterization, pp. 117–134 ACS Symp. Series 28 R (Amer. chem Soc. 1976).

HALL, L.L.; SMITH, F.A.; DELOPEZ, O., and GARDNER, D.E.: Direct potentiometric determination of total ionic fluoride in biological fluids. Clin. Chem. *18:* 1455–1458 (1972).

HAM, M.P. and SMITH, M.D.: Fluorine balance studies on three women. J. Nutr. *53:* 225–232 (1954).

HANHIJÄRVI, H.: Comparison of free ionized fluoride concentrations of plasma and renal clearance in patients of artificially fluoridated and non-fluoridated drinking water areas. Finn. dent. Soc. Proc., suppl. *3* (1974).

HARGREAVES, J.A.; INGRAM, G.S., and WAGG, B.J.: Excretion Studies on the ingestion of a monofluorophosphate toothpaste by children. Caries Res. *4:* 256–268 (1970).

HARGREAVES, J.A.; INGRAM, G.S., and WAGG, B.J.: A gravimetric study of the ingestion of toothpaste by children. Caries Res. *6:* 237–243 (1972).

HARKINS, R.W.; LONGENECKER, J.B., and SARETT, H.P.: Effect of sodium fluoride on the growth of rats with varying vitamin and calcium intakes. J. Nutr. *81:* 81–86 (1963).

HARRISON, M.F.: Fluorine content of teas consumed in New Zealand. Br. J. Nutr. *3:* 162–166 (1949).

HASSELMAN, G. and ROHOLT, K.: A note on the stability of the trifluoromethyl group of bendroflumethiazide in rats. J. Pharm. Pharmac. *15:* 339–340 (1963).

HELLSTRØM, I.: Fluorine retention following sodium fluoride mouthwash. Acta odont. scand. *18:* 263–278 (1960).

HELLSTRØM, I. and ERICSSON, Y.: Urinary fluoride excretion in small children following short term fluoride supply with tablets or domestic salt. Scand. J. dent. Res. *84:* 187–199 (1976).

HENNON, D.K.; STOOKEY, G.K., and MUHLER, J.C.: Blood and urinary fluoride levels associated with ingestion of sodium fluoride containing vitamin tablets. J. dent. Res. *48:* 1211–1215 (1969).

HENSCHLER, D.; BUTTNER, W., and PATZ, J.: Absorption, distribution in body fluids and bioavailability of fluoride; in KUHLENCORDT and KRUSE Calcium metabolism, bone and metabolic bone diseases, pp. 111–121 (Springer, Berlin 1975).

HILL, T.J.: Oral pathology, p. 37 (Lea & Febiger, Philadelphia 1945).

HODGE, H.C. and SMITH, F.A.: Occupational fluoride exposure. J. occ. Med. *19:* 12–39 (1977).

HORIKAWA, E.K.; STEELMAN, R.; FLANAGAN, T.L., and VAN LOON, E.J.: Determination of fluoride in femurs of rats receiving organic fluoride. J. med. pharm. Chem. *2:* 541–551 (1960).

JARDILIER, J.C. et DESMET, G.: Etude du fluor sérique et de ses combinaisons par une technique utilisant une électrode spécifique. Clinica chim. Acta *47:* 357–363 (1973).

JONES, C.M.; HARRIES, J.M., and MARTIN, A.E.: Fluorine in leafy vegetables. J. Sci. Fd Agric. *22:* 602–605 (1971).

JOWSEY, J.; JOHNSON, W.J.; TAVES, D.R., and KELLY, P.J.: Effects of dialysate Ca and fluoride on bone disease during regular hemodialysis. J. Lab. clin. Med. *79:* 204–213 (1972).

KALLIS, D.G. and SILVA, D.G.: Carnavon studies. III. Detailed investigations related to endemic fluorosis present in children in Carnavon, Western Australia (Aug. 1963). Aust. dent. J. *15:* 35–43 (1970).

KOCH, G. and FRIBERGER, P.: Fluoride content of outermost enamel layer in teeth exposed to topical fluoride applications. Odont. Rev. *22:* 351–362 (1971).

KRAMER, L.; OSIS, D.; WIATROWSKI, E., and SPENCER, H.: Dietary fluoride in different areas in the US. Am. J. clin. Nutr. *27:* 590–594 (1974).

LARGENT, E.J. and HEYROTH, F.F.: The absorption and excretion of fluorides .III. Further observations on metabolism of fluorides at high levels of intake. J. Ind. Hyg. Toxicol. *31:* 134 (1949).

LARGENT, E.J.: Fluorosis, pp. 22–57 (Ohio State Univ. Press, Columbus 1961).

LEATHERWOOD, E.C.; BURNETT, G.W.; CHANDRAVEJJSMARN, R., and SIRIKAYA, R.: Dental caries and dental fluorosis in Thailand. Am. J. Publ. Hlth *55:* 1792-1799 (1965).

LEE, J.R.: Optimal fluoridation. The concept and its application to municipal water fluoridation. West. J. Med. *122:* 431–436 (1975).

LEX, C.: Fluoride content of teeth in endemic fluorosis. Dt. zahnärztl. Z. *29:* 791–794 (1974).

LIKINS, R.C.; McCLURE, F.J., and STEERE, A.C.: Urinary excretion of fluoride following defluoridation of a water supply. Publ. Hlth Rep. ,Wash. *71:* 217–220 (1956).

LONG, C.: Biochemists handbook, p. 1119 (Van Nostrand, Princeton 1961).

LOUGH, J.; NOONAN, R.; GAGNON, R., and KAYE, M.: Effects of fluoride on bone in chronic renal failure. Archs Path. *99:* 484–487 (1975).

MACHLE, W.; SCOTT, E.W., and LARGENT, E.J.: The absorption and excretion of fluorides.

I. The normal fluoride balance. J. Ind. Hyg. Toxicol. *24:* 199–204 (1942).

MADDOCK, R.K., jr.: Fluorosis, fluid fluoridation and food fluorine. Ann. intern. Med. *70:* 1049–1050 (1969).

MARGOLIS, F.J.; NAGLER, R.C., and HOLKEBOER, P.E.: Short term fluoride excretion in young children. Am. J. Dis. Child. *113:* 673–676 (1967).

MARIER, J.R. and ROSE, D.: The fluoride content of some foods and beverages. A brief survey using a modified Zr-Spadns method. J. Fd Sci. *31:* 941–946 (1966).

MASSLER, M. and SCHOUR, I.: Relation of endemic fluorosis to malnutrition. J. Am. dent. Ass. *44:* 156–169 (1952).

McCLURE, F.J.: Fluorides in food and drinking water. Natn Inst. Hlth Bull. *172:* (1939).

McCLURE, F.J.: Ingestion of fluoride and dental caries. Am. J. Dis. Child. *66:* 362–369 (1943).

McCLURE, F.J.: Fluorine in foods. Survey of recent data. Publ. Hlth Rep., Wash. *64:* 1061 (1949).

McCLURE, F.J.: The availability of fluorine in sodium fluoride vs. sodium fluorsilicate. Publ. Hlth Rep., Wash. *65:* 1175–1186 (1950).

McCLURE, F.J. and KINSER, C.A.: Fluoride domestic waters and systemic effects. II. Fluorine content of urine in relation to fluorine in drinking water. Publ. Hlth Rep., Wash. *59:* 1575–1591 (1944).

McCLURE, F.J.; MITCHELL, H.H.; HAMILTON, T.S., and KINSER, C.A.: Balances of fluorine ingested from various sources in food and water by five young men. Excretion thru the skin. J. Ind. Hyg. Toxicol. *27:* 159–170 (1945).

McPHAIL, C.W.B. and ZACHERL, W.: Fluid intake and climatic temperature. Relation to fluoridation. J. Can. dent. Ass. *31:* 7–16 (1965).

MELLANDER, O.; VAHLQUIST, B., and MELLBIN, T.: Breast feeding and artificial feeding. A serological and biolchemical study in 402 infants with a survey of the literature. Acta paedt. scand., suppl. 116 (1959).

MILGALTER, N.; ZADIK, D.; GEDALIA, I., and KELMAN, M.: Fluorosis and dental caries in the region of Jotvata Israel. J. dent. med. *23:* 104–109 (1974).

MINOGUCHI, G.: Japanese studies on water and food fluoride and general dental health; in Fluorides and human health, pp. 294–304 (WHO, Geneva 1970).

MØLLER, I.J.; PINDBORG, J.; GEDALIA, I., and ROED-PETERSEN, B.: The prevalence of dental fluorosis in the people of Uganda. Archs oral. Biol. *15:* 213–225 (1970).

MØLLER, I.J. and POULSEN, S.: A study of dental mottling in children in Khouribga, Morocco. Archs oral Biol. *20:* 601–607 (1976).

MØLLER, K.J.; SCHAIT, A., and MUHLEMANN, H.R.: Fluorinholdet I den superfichielle del al fluobehandlede of ikke-fluorbehandlede taenders emalje. Tandlaegebl. *69:* 849–862 (1965).

MURRAY, M.M.; FORREST, J.R.; GRIFFITH, G.W., and LONGWELL, J.: Iodine and fluorine nutrition. Nature, Lond. *177:* 912–914 (1956).

MURRAY, M.M. and WILSON, D.C.: Fluorosis and nutrition in Morocco. Br. dent. J. *84:* 97–100 (1948).

MYERS, H.M.; HAMILTON, J.G., and BECKS, H.: A tracer study of the transfer of ^{18}F to teeth by topical application. J. dent. Res. *31:* 743–746 (1952).

NANDA, R.S.; ZIPKIN, I.; DOYLE, J., and HOROWITZ, H.S.: Factors affecting the prevalance of dental fluorosis in Lucknow, India. Archs oral Biol. *19:* 781–792 (1974).

National Academy of Sciences: Fluorides, biologic effects of atmospheric pollutants (Division Med. Sci. Natn. Res. Council, Washington 1971).

OELSCHLAGER, W.: Fluoride in food. Fluoride Qt. Rep. *3:* 6–11 (1970).

OKAMURA, T. and MATSUSHIDA, T.: Fluoride intake in Japan. Chem. Abstr. *70:* No. 104101 (1969).

OREOPOULOS, D.G.; TAVES, D.R.; RABINOVICH, S.; MEEMA, H.E.; MURRAY, T.; FENTON, S.S., and VEBER, DE G.A.: Fluoride and dialysis osteodystrophy. Results of a double blind study. Trans. Am. Soc. artif. internal Organs *20:* 203–208 (1974).

OSIS, D.; KRAMER, L.; WIATROWSKI, E., and SPENCER, H.: Dietary fluoride intake in man. J. Nutr. *104:* 1313–1318 (1974a).

OSIS, D.; WIATROWSKI, E.; SAMACHSON, J., and SPENCER, H.: Fluoride analysis of the human diet and of biological samples. Clinica chim. Acta *51:* 211–216 (1974b).

PARKINS, F.M.; TINANOFF, N.; MOUTINHO, M.; ANSTEY, M.B., and WUZIRI, M.H.: Relationships of human plasma fluoride and bone fluoride to age. Calcif. Tiss. Res. *16:* 335–338 (1974).

POSEN, G.A.; GRAY, D.G.; JAWORSKI, Z.F.; COUTURE, R., and RASHID, A.: Comparison of renal osteodystrophy in patients dialyzed with deionized and non-deionized water. Trans. Am. Soc. artif. internal Organs *18:* 405–411 (1972).

POT, T.J. and FLISSEBAALJE, T.D.: Fluoride uptake excretion and serum concentrations among foundry workers in an artificially fluoridated region. Ned. Tijdschr. Tandheelk. *81:* 430–435 (1974).

POULSEN, S. and MØLLER, I.J.: Gingivitis and dental plaque in relation to dental fluorosis in man in Morocco. Archs oral Biol. *91:* 951–954 (1974).

PU, M.Y. and LILIENTHAL, B.: Dental caries and mottled enamel among Formosan children. Archs oral Biol. *5:* 125–136 (1961).

RAMSEY, A.C.; HARDWICK, J.L., and TAMACAS, J.C.: Fluoride intakes and caries increments in relation to tea consumption by British children. Caries Res. *9:* 312–314 (1975).

REDDY, G.S. and NARASINGA RAO, B.S.: Ca turnover with calcium and phosphorus balances. Metabolism *20:* 650–656 (1971).

REDDY, G.S. and SRIKANTIA, S.G.: Effect of dietary calcium, vitamin C and protein in development of experimental skeletal fluorosis. Metabolism *20:* 642–649 (1971).

RICHARDS, L.F.; WESTMORELAND, W.W.; TASHIRO, M.; McKAY, C.H., and MORRISON, J.T.: Determining optimum fluoride levels for community water supplies in relation to temperature. J. Am. dent. Ass. *74:* 389–397 (1967).

RUZICKA, J.A.; MRKLAS, L., and ROKYTOVA, K.: Influence of water intake on the degree of incisor fluorosis and on the incorporation of fluoride into bones and incisor teeth of mice. Caries Res. *7:* 166–172 (1973).

RUZICKA, J.A.; MRKLAS, L., and ROKYTOVA, K.: Incorporation of fluoride in bones and teeth and dental fluorosis in mice after administration of complex fluorides. Archs Oral Biol. *19:* 947–950 (1974).

SANFILIPPO, F.A. and BATTISTONE, G.C.: The fluoride content of a representative diet of the young adult male. Clinic. chim. Acta *31:* 453–457 (1971).

SAUERBRUNN, B.J.L.; RYAN, C.M., and SHAW, J.F.: Chronic fluoride intoxication with fluorotic radiculomyelopathy. Ann. Intern. Med. *63:* 1074–1078 (1965).

SINGER, L.: Fluoride metabolism, chap. 1; in PICOZZI and SMUDSKI Pharmacology of fluorides. 12th Symp. IADR Pharmac., Ther. and Toxicol Group, pp. 5–14 1974.

SINGER, L. and ARMSTRONG, W.D.: Regulation of human plasma fluoride concentration. J. appl. Physiol. *15:* 508–510 (1960).

SINGER, L.; OPHAUG, R.H., and ARMSTRONG, W.D.: Influence of variations in fluoride intake on the ionic and bound fractions of plasma and muscle fluoride. Proc. Soc. exp. Biol. Med. *151:* 627–631 (1976).

SMITH, F.A.: Biologic properties of selected fluorine-containing organic compounds, chap. 7; in SMITH Pharmacology of fluorides, pp. 253–389 (Springer, Berlin 1970).

SPENCER, H.; KRAMER, L.; OSIS, D., and WIATROWSKI, E.: Excretion of retained fluoride in man. J. appl. Physiol. *38:* 282–287 (1975).

SPENCER, H.; LEWIN, I.; FOWLER, J., and SAMACHSON, J.: Effect of NaF on calcium absorption and balance in man. Am. J. clin. Nutr. *22:* 381–390 (1969).

SPENCER, H.; OSIS, D.; WIATROWSKI, E., and SAMACHSON, J.: Availability of fluoride from fish protein concentrate and from NaF in man. J. Nutr. *100:* 1415–1424 (1970).

STOOKEY, G.K.: Ingestion from drugs, chap. 2, sect. 4; in Fluorides and human health, pp. 48–49 (WHO, Geneva 1970).

SUTTIE, J.W.: Effects of fluoride on livestock. J. occ. Med. *19:* 40–48 (1977).

SUTTIE, J.W.; CARLSON, J.R., and FALTIN, E.C.: Effect of alternating periods of high and low fluoride ingestion in dairy cows. J. Dairy Sci. *55:* 790–804 (1972).

TAVES, D.R.: Normal human fluoride concentrations. Nature, Lond. *211:* 192–193 (1966).

TAVES, D.R.; TERRY, R.; SMITH, F.A., and GARDNER, D.E.: Use of fluoridated water in long term hemodialysis. Archs intern Med. *115:* 167–172 (1965).

TORELL, P. and ERICSSON, Y.: Two year clinical tests with different methods of local caries preventive fluorine application in Swedish school children. Acta odont. scand. *23:* 287–322 (1965).

TOTH, K.: Fluoride ingestion related to body weight. Caries Res. *9:* 290–291 (1975).

VANDYKE, R.A. and WOOD, C.L.: Metabolism of methoxyflurane. Release of inorganic fluoride in human and rat hepatic microsomes. Anesthesiology *39:* 613–618 (1973).

VAN RENSBURG, B.G.J.: Protein deficient diet, fluoride and amelogenesis. J. dent. Ass. S. Afr. *27:* 204–209 (1972).

VENKATESWARLU, P.: Determination of total fluorine in serum and other biological materials by oxygen bomb and reverse extraction techniques. Analyt. Biochem. *68:* 512–521 (1975).

WAGNER, M.J.: A comparison of fluorides as they naturally occur and as they are added in communal fluoridation, chap. 2; in MUHLER and M.K. HINE Fluorine and dental health pp. 38–59 (Univ. of Indiana Press, Bloomington 1959).

WALKER, A.R.P.: The human requirement of calcium. Should low intakes be supplemented. Am. J. clin. Nutr. *25:* 518–530 (1972).

WALKER, J.S.; MARGOLIS, F.J.; LUTEN, T.H.; WEIL, M.L., and WILSON, H.L.: Water intake of normal children. Science *140:* 890 (1963).

WEATHERELL, J.A.; DEUTSCH, D.; ROBINSON, C., and HALLSWORTH, A.S.: Assimilation of fluoride by enamel throughout the life of the tooth. Caries Res. *11:* suppl. 1, pp. 85–115 (1977).

WEDDLE, D.A. and MUHLER, J.C.: Effects of Inorganic salts on fluorine storage in rats. J. Nutr. *54:* 437–444 (1954).

WEINSTEIN, L.H.: Fluoride and plant life. J. occ. Med. *19:* 49–78 (1977).

WIATROWSKI, E.; KRAMER, L.; OSIS, D., and SPENCER, H.: Dietary fluoride intake of infants. Pediatrics, Springfield *55:* 517–522 (1975).

WIDGER, L.A.; GANDOLFI, A.J., and VAN DYKE, R.A.: Hypoxia and halothane metabolism *in vivo*. Release of inorganic fluoride and halothane metabolite binding to
  cellular constituents. Anesthesiology *44:* 197–201 (1976).
WIRZ, R.: Ergebnisse des Grossversuches mit fluoridierter Milch in Winterthur von 1958
  bis 1964. Schweiz. Mschr. Zahnheilk. *74:* 767–784 (1964).
ZIPKIN, I.; LIKINS, R.C., and MCCLURE, F.J.: Deposition of fluoride, calcium and phosphorus in experimental low phosphorus rickets. J. Nutr. *67:* 59–68 (1959).
ZIPKIN, I.; LIKINS, R.C.; MCCLURE, F.J., and STEERE, A.C.: Urinary fluoride levels
  associated with the use of fluoridated drinking water. Publ. Hlth Rep., Wash. *71:*
  767–772 (1956).
ZIPKIN, I.; MCCLURE, F.J.; LEONE, N.C., and LEE, W.A.: Fluoride deposition in human
  bones after prolonged ingestion of fluoride in drinking water. Publ. Hlth Rep.,
  Wash. *73:* 732–740 (1958).
ZIPKIN, I.; MCCLURE, F.J., and LEE, W.A.: Relation of the fluoride content of human
  bone to its chemical composition. Archs oral. Biol. *2:* 190–195 (1960).

Subject Index